Computational Electromagnetics for Industrial Applications

Computational Electromagnetics for Industrial Applications

Editors

Giulio Antonini
Daniele Romano
Luigi Lombardi

Basel • Beijing • Wuhan • Barcelona • Belgrade • Novi Sad • Cluj • Manchester

Editors

Giulio Antonini	Daniele Romano	Luigi Lombardi
University of L'Aquila	University of L'Aquila	Micron Semiconductor
L'Aquila	L'Aquila	Avezzano
Italy	Italy	Italy

Editorial Office
MDPI
St. Alban-Anlage 66
4052 Basel, Switzerland

This is a reprint of articles from the Special Issue published online in the open access journal *Electronics* (ISSN 2079-9292) (available at: https://www.mdpi.com/journal/electronics/special_issues/CEIA_electronics).

For citation purposes, cite each article independently as indicated on the article page online and as indicated below:

Lastname, A.A.; Lastname, B.B. Article Title. *Journal Name* **Year**, *Volume Number*, Page Range.

ISBN 978-3-0365-8732-5 (Hbk)
ISBN 978-3-0365-8733-2 (PDF)
doi.org/10.3390/books978-3-0365-8733-2

Contents

About the Editors

Giulio Antonini

Giulio Antonini received the Laurea degree cum laude in electrical engineering from the University of L'Aquila, L'Aquila, Italy, in 1994 and the Ph.D. degree in electrical engineering from University of Rome "La Sapienza" in 1998. Since 1998, he has been with the UAq EMC Laboratory, University of L'Aquila, where he is currently a professor. His scientific interests are in the field of computational electromagnetics. He has authored more than 300 papers published on international journals and in the proceedings of international conferences. He has also co-authored the book "Circuit Oriented Electromagnetic Modeling Using the PEEC Techniques", Wiley–IEEE Press, 2017.

Daniele Romano

Daniele Romano was born in Campobasso, Italy, in 1984. He received the Laurea degree in computer science and automation engineering in 2012 from the University of L'Aquila, L'Aquila, Italy. He received the Ph.D. degree in 2018. Since 2012, he has been with the UAq EMC Laboratory, University of L'Aquila, focusing on EMC modeling and analysis, algorithm engineering and speed-up techniques applied to EMC problems.

Luigi Lombardi

Luigi Lombardi received the Laurea (MD) in electronic engineering and the PhD, both cum laude, from the University of L'Aquila, L'Aquila, Italy. In early 2018, he was a visiting researcher with the department of electronics in Carleton University, Ottawa, Ontario, Canada. Since November 2018, he has been working as a design engineer with the Non-Volatile Memory (NVM) design team at Micron Semiconductor, Avezzano, AQ, Italy.

 *electronics*

Editorial

Computational Electromagnetics for Industrial Applications

Giulio Antonini [1,*], Daniele Romano [1] and Luigi Lombardi [2]

[1] Department of Industrial and Information Engineering and Economy, University of L'Aquila, 67100 L'Aquila, Italy; daniele.romano@univaq.it

[2] Micron Semiconductor, 67051 Avezzano, Italy; luigilombardi89@gmail.com

* Correspondence: giulio.antonini@univaq.it

1. Introduction

Nowadays, computational electromagnetics (CEMs) methods play an important role in the rapid modeling and design of electromagnetic (EM) systems and their industrial applications. Virtual prototyping based on computational electromagnetics is currently widely adopted in electrical and electronic systems design because of the high accuracy guaranteed by many numerical methods for the solution of Maxwell's equations in a wide range of frequency from DC to hundreds of GHz or even in the THz range. With the continuous increase of integration and complexity in EM systems and devices, numerical modeling and simulation methods play a key role in the design of electromagnetic systems. From this perspective, accurate and efficient algorithms and methods enabling an accurate and efficient analysis of complex EM problems are strongly requested. Furthermore, semi-analytical methods can also offer elegant and accurate solutions to complex EM problems. This special issue aims to promote cutting-edge research along this direction and offers a timely collection of works to benefit researchers and practitioners.

2. The Special Issue

This Special Issue consists of nine papers covering a broad range of topics related to the applications of EM waves, from the modeling of transmission lines and electromagnetic structures to imaging, from the high frequency characterization of wave guides and meta-surfaces, to machine-learning modeling of electromagnetic devices and field distribution along the earth's surface of a submerged line current at low frequency. The contents of these papers are briefly introduced in the following.

In the first paper [1], an analytical series-form solution for the time-harmonic EM field components produced by an overhead current line source is presented. The axial electric field is expanded into a power series of the vertical propagation coefficient in the air space leading to explicit expressions for the electric and magnetic fields.

The second paper [2] presents an extension of the formulation of wave propagation in transverse electric (TE) and transverse magnetic (TM) modes for the case of metallic cylindrical waveguides filled with longitudinally magnetized ferrite which are important for various industrial applications, such as circulators, isolators, antennas and filters. The proposed approach guarantees a considerable reduction of the computation time compared to a commercial software.

In the third paper [3], the addition translation theorem is used to analyze multiple scattering by two Perfect Electric Conducting (PEC) spheres. The proposed approach will allow to evaluate the scattering from macro-structures composed of spherical particles, i.e., biological molecules, clouds of airborne particles.

The fourth paper [4] presents a simple technique to identify material texture from afar, by using polarization-resolved imaging. It has been applied to both isotropic references (Teflon bar) and anisotropic samples (wood). Such a technique can be easily implemented into industrial environments, where fast and cheap sensors are required.

Citation: Antonini, G.; Romano, D.; Lombardi, L. Computational Electromagnetics for Industrial Applications. *Electronics* **2022**, *11*, 1830. https://doi.org/10.3390/electronics11121830

Received: 6 June 2022

Accepted: 8 June 2022

Published: 9 June 2022

Publisher's Note: MDPI stays neutral with regard to jurisdictional claims in published maps and institutional affiliations.

The EM scattering may be a significant source of degradation in signal and power integrity of high-contrast silicon-on-insulator (SOI) nano-scale interconnects, such as opto-electronic or optical interconnects operating at 100 s of THz. The fifth paper [5] characterizes THz scattering loss in nano-scale SOI waveguides exhibiting stochastic surface roughness with exponential autocorrelation.

Paper [6] considers the calculation of magnetic flux density in the vicinity of overhead distribution lines which takes the higher current harmonics into account. Based on the Biot–Savart law and the complex image method, it allows to determine the contributions of individual harmonic components of the current intensity to the total value of magnetic flux density.

A machine learning (ML)-based regression for the construction of complex-valued surrogate models for the analysis of the frequency-domain responses of EM structures is presented in the seventh paper [7]. Relying on the combination of the principal component analysis (PCA) and an unusual complex-valued formulation of the Least Squares Support Vector Machine (LS-SVM) regression, a compact set of complex-valued surrogate models are trained and then used to obtain the frequency response of EM structures.

Dielectric metasurfaces have emerged as a promising technology for sensing applications and non-linear frequency conversion, and for their flexibility to shape the emission pattern in the visible regime. In paper [8], finite-size metasurfaces and beam-like illumination conditions are investigated in contrast to the typical infinite plane-wave illumination compatible with the Floquet theorem.

Finally, a numerical analysis on field distribution along the earth's surface of a line current source submerged in the ground is conducted in [9]. The potential of the extremely low frequency (ELF) technology in the envisioned long-distance communication techniques is investigated by the combined numerical methods of the Romberg–Euler method and Gauss–Laguerre method.

Conflicts of Interest: The authors declare no conflict of interest.

References

1. Parise, M. On the Electromagnetic Field of an Overhead Line Current Source. *Electronics* **2020**, *9*, 2009. [CrossRef]
2. Sakli, H. Cylindrical Waveguide on Ferrite Substrate Controlled by Externally Applied Magnetic Field. *Electronics* **2021**, *10*, 474. [CrossRef]
3. Batool, S.; Nisar, M.; Dinia, L.; Mangini, F.; Frezza, F. Multiple Scattering by Two PEC Spheres Using Translation Addition Theorem. *Electronics* **2022**, *11*, 126. [CrossRef]
4. Fazio, E.; Batool, S.; Nisar, M.; Alonzo, M.; Frezza, F. Recognition of Bio-Structural Anisotropy by Polarization Resolved Imaging. *Electronics* **2022**, *11*, 255. [CrossRef]
5. Guiana, B.; Zadehgol, A. Characterizing THz Scattering Loss in Nano-Scale SOI Waveguides Exhibiting Stochastic Surface Roughness with Exponential Autocorrelation. *Electronics* **2022**, *11*, 307. [CrossRef]
6. Mujezinović, A.; Turajlić, E.; Alihodžić, A.; Dedović, M.M.; Dautbašić, N. Calculation of Magnetic Flux Density Harmonics in the Vicinity of Overhead Lines. *Electronics* **2022**, *11*, 512. [CrossRef]
7. Soleimani, N.; Trinchero, R. Compressed Complex-Valued Least Squares Support Vector Machine Regression for Modeling of the Frequency-Domain Responses of Electromagnetic Structures. *Electronics* **2022**, *11*, 551. [CrossRef]
8. Ciarella, L.; Tognazzi, A.; Mangini, F.; De Angelis, C.; Pattelli, L.; Frezza, F. Finite-Size and Illumination Conditions Effects in All-Dielectric Metasurfaces. *Electronics* **2022**, *11*, 1017. [CrossRef]
9. Yang, K.; Wang, J.; Liu, S.; Ding, K.; Li, H.; Li, B. Investigations on Field Distribution along the Earth's Surface of a Submerged Line Current Source Working at Extremely Low Frequency Band. *Electronics* **2022**, *11*, 1116. [CrossRef]

 *electronics*

Article

On the Electromagnetic Field of an Overhead Line Current Source

Mauro Parise

Faculty of Engineering, University Campus Bio-Medico of Rome, Via Alvaro del Portillo 21, 00128 Rome, Italy; m.parise@unicampus.it

Received: 8 October 2020; Accepted: 25 November 2020; Published: 27 November 2020

Abstract: This work presents an analytical series-form solution for the time-harmonic electromagnetic (EM) field components produced by an overhead current line source. The solution arises from casting the integral term of the complete representation for the generated axial electric field into a form where the non-analytic part of the integrand is expanded into a power series of the vertical propagation coefficient in the air space. This makes it possible to express the electric field as a sum of derivatives of the Sommerfeld integral describing the primary field, whose explicit form is known. As a result, the electric field is given as a sum of cylindrical Hankel functions, with coefficients depending on the position of the field point relative to the line source and its ideal image. Analogous explicit expressions for the magnetic field components are obtained by applying Faraday's law. The results from numerical simulations show that the derived analytical solution offers advantages in terms of time cost with respect to conventional numerical schemes used for computing Sommerfeld-type integrals.

Keywords: line current source; half-space problem; EM wave propagation

1. Introduction

The computation of the electromagnetic (EM) fields from overhead electric lines located above dissipative terrestrial areas is a classical problem which is still of interest today because of the public concern about the biological effects of field exposure. In fact, it is well known that the strong EM fields associated with the high-intensity currents and voltages of power lines can generate undesirable effects on humans, animals and other forms of life [1–3]. The problem, in all of its variants, has spawned a number of contributions to the scientific literature since a solution was first developed at the beginning of the 20th century by Carson [1–25]. Yet, in spite of the variety of proposed approaches, to Author's knowledge there is scarcity of purely analytical techniques in literature, and most of the published methods attempt to solve the problem through the derivation of analytical formulations amenable to numerical treatment. Excellent illustrations of these approaches are the well-known method of moments (MoM), which is used to solve surface integral formulations, and the method of auxiliary sources [10], consisting of expressing the fields as weighted superpositions of the contributions from a finite number of fictitious currents flowing on mathematical surfaces separated from the air-ground interface. Even if these methods may exhibit good performances in terms of accuracy and efficiency, as numerical procedures they suffer from the intrinsic disadvantage of involving memory requirements and computational costs significantly greater than those implied by an analytical solution. Moreover, they provide less insight in the physics of the problem and are also less suitable for sensitivity analysis. On the other hand, one important contribution in the context of purely analytical techniques is the work by Prof. J. R. Wait [5], which presents field expressions derived through usage of the complex image theory. The only drawback of the obtained formulas resides in that they are valid in the quasi-static regime only, that is when the effects of the displacement currents in the air space are negligible. The aim of the present paper is to derive rigorous series-form expressions for the time-harmonic

EM field components produced in the air-space by a uniform-current line source located above a homogeneous dissipative ground. The field expressions are obtained starting from decomposing the integral representation for the axial component of the generated electric field into three parts, that is the direct field induced by the source current, the ideal reflected field induced by a negative image current and a correction term due to the imperfect conductivity of the ground. Next, after recognizing that the first two terms can be straightforwardly expressed in explicit form, the non-analytic part of the integrand of the correction term is expanded into a power series of the vertical propagation coefficient in the air space. This permits to express the electric field as a sum of derivatives of the Sommerfeld Integral describing the direct field. As a result, the field is given as a sum of cylindrical Hankel functions, with coefficients depending on the position of the field point relative to the original current source and its ideal image. Finally, analogous explicit expressions for the magnetic field components are obtained by applying Faraday's law. The obtained solution is not subject to simplifying assumptions, and hence is valid even when the high-frequency effects due to the displacement currents in both the air and the soil are not negligible. As a consequence, the solution is also of practical use as an analytical benchmark for simulation tools employed to solve EM boundary value problems. Numerical tests are performed to show the validity of the developed method and its advantages in terms of computation time with respect to standard numerical algorithms used to evaluate Sommerfeld-type integrals.

2. Formulation of the Problem

Consider an infinite line source of current located in proximity of a flat homogeneous ground, as shown in Figure 1.

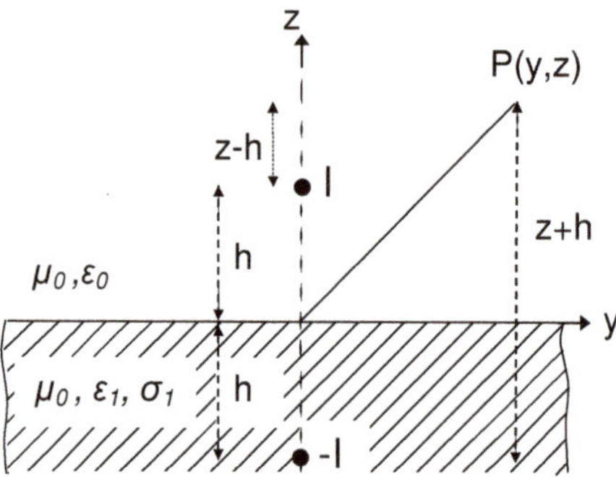

Figure 1. Sketch of a line source of current above a homogeneous ground.

The line is situated above the ground at height h, and the material medium is characterized by the dielectric permittivity ϵ_1 and electrical conductivity σ_1, while the magnetic permeability is taken to be everywhere that of free-space μ_0. Under the hypothesis that the observation point lies in the near-field region of the line source, it is reasonable to assume that the line supports a nearly uniform current [1], i.e., $I(x,t) \cong Ie^{j\omega t}$. Due to the symmetry of the problem, the total electric field generated in the air space has only one component in the axial direction, whose integral representation is well known and given by [10]

$$E_x = -\frac{j\omega\mu_0 I}{2\pi}\left[S_d\left(|z-h|\right) + S_r\left(z+h\right)\right],\tag{1}$$

where

$$S_d(\zeta) = \int_0^\infty \frac{1}{u_0}e^{-u_0\zeta}\cos\left(\lambda y\right)d\lambda = -\frac{j\pi}{2}H_0^{(2)}\left(k_0\sqrt{y^2+\zeta^2}\right)\tag{2}$$

and

$$S_r(\zeta) = \int_0^\infty \frac{1}{u_0} \frac{u_0 - u_1}{u_0 + u_1} e^{-u_0\zeta} \cos(\lambda y)\, d\lambda \tag{3}$$

are, respectively, the Sommerfeld Integrals describing the direct and reflected fields, being $H_0^{(2)}$ the zeroth-order Hankel function of the second kind, and $u_n = \sqrt{\lambda^2 - k_n^2}$. The aim of this paper is to derive exact series representations for E_x and the magnetic field components, the latter arising from applying Faraday's law. To this goal, it is first convenient to use the identity

$$\frac{u_0 - u_1}{u_0 + u_1} = -1 + \frac{2u_0}{u_0 + u_1} \tag{4}$$

and split S_r into two terms, that is the ideal reflected field contribution (induced by a negative image current) plus a term due to the imperfect conductivity of the ground. It reads

$$S_r\,(z+h) = -\,S_d\,(z+h) + S_c\,(z+h)\,, \tag{5}$$

with

$$S_c\,(\zeta) = 2 \int_0^\infty \frac{e^{-u_0\zeta}}{u_0 + u_1} \cos(\lambda y)\, d\lambda, \tag{6}$$

and where account has been taken of (2). The integral S_c may be evaluated by proceeding as follows. After setting $\alpha = \sqrt{k_1^2/k_0^2 - 1}$ and

$$\frac{1}{u_0 + u_1} = \frac{u_0}{k_0^2\alpha^2} + jP(u_0), \tag{7}$$

we replace the quantity $e^{u_0\beta}P(u_0)$, where β is a non-negative real constant to be determined, with its Taylor expansion about $u_0 = 0$, in a similar fashion as in [5]. It yields

$$f(u_0) = e^{u_0\beta}P(u_0) = \sum_{i=0}^\infty \frac{f^{(i)}(0)}{i!} u_0^i, \tag{8}$$

and, as a consequence, (6) becomes

$$S_c\,(\zeta) = \frac{2}{(k_0\alpha)^2} \int_0^\infty u_0 e^{-u_0\zeta} \cos(\lambda y)\, d\lambda + 2j \sum_{i=0}^\infty a_i(\beta) \int_0^\infty u_0^i e^{-u_0(\zeta+\beta)} \cos(\lambda y)\, d\lambda, \tag{9}$$

being $a_i(\beta) = f^{(i)}(0)/i!$. Under the assumption that $\zeta > 0$, the second integral on the right-hand side of (9) converges regardless of the value chosen for the non-negative arbitrary constant β. This happens because, as λ becomes sufficiently large, the exponential factor $e^{-u_0(\zeta+\beta)}$ approaches $e^{-\lambda(\zeta+\beta)}$, which rapidly decays with increasing λ. Since β may be assumed to be as small as desired, we now take the limit of (9) as $\beta \to 0^+$. It is not difficult to prove that, subject to this condition, the odd coefficients a_{2i+1} ($i = 0, 1, 2, \ldots$) become null. On the other hand, the limit of the even coefficients reads

$$\lim_{\beta \to 0^+} a_{2i}(\beta) = \frac{c_{2i}}{(k_0\alpha)^{2i+1}}, \qquad \text{with} \quad c_i = \frac{(i+1)!!}{(i^2 - 1)\, i!!}, \tag{10}$$

and, as a consequence, (9) may be expressed as

$$S_c\,(\zeta) = \frac{2}{(k_0\alpha)^2} \frac{\partial^2 S_d}{\partial \zeta^2} - 2j \sum_{i=0}^\infty \frac{c_{2i}}{(k_0\alpha)^{2i+1}} \frac{\partial^{2i+1} S_d}{\partial \zeta^{2i+1}}, \tag{11}$$

where the ζ-derivatives of S_d, that is the derivatives of $H_0^{(2)}$, are to be made explicit. After letting let $r = \sqrt{y^2+\zeta^2}$, from the differential properties of the Bessel functions [26] it follows that

$$\frac{\partial H_0^{(2)}}{\partial \zeta} = -\frac{k_0\,\zeta}{r}H_1^{(2)}$$

$$\frac{\partial^2 H_0^{(2)}}{\partial \zeta^2} = -k_0\left[1+\zeta^2\left(\frac{1}{r}\frac{d}{dr}\right)\right]\frac{H_1^{(2)}}{r}$$

$$\frac{\partial^3 H_0^{(2)}}{\partial \zeta^3} = -k_0\left[3\zeta\left(\frac{1}{r}\frac{d}{dr}\right)+\zeta^3\left(\frac{1}{r}\frac{d}{dr}\right)^2\right]\frac{H_1^{(2)}}{r}$$

$$\frac{\partial^4 H_0^{(2)}}{\partial \zeta^4} = -k_0\left[3\left(\frac{1}{r}\frac{d}{dr}\right)+6\zeta^2\left(\frac{1}{r}\frac{d}{dr}\right)^2+\zeta^4\left(\frac{1}{r}\frac{d}{dr}\right)^3\right]\frac{H_1^{(2)}}{r}$$

$$\vdots$$

$$\frac{\partial^l H_0^{(2)}}{\partial \zeta^l} = -k_0\sum_{m=0}^{\lfloor l/2\rfloor}(-1)^m d_{lm}\zeta^{l-2m}\left(\frac{1}{r}\frac{d}{dr}\right)^{l-m-1}\frac{H_1^{(2)}}{r}, \tag{12}$$

where the argument of the Hankel functions, that is $k_0 r$, has been omitted for notational simplicity, and with

$$d_{lm} = (-1)^m\binom{l}{l-2m}(2m-1)!! = \frac{(-1)^m l!}{(l-2m)!(2m)!!}. \tag{13}$$

It should be noted that the number of terms on the right-hand sides of (12) is $(l/2+1)$ for even l, and $[(l-1)/2+1]$ for odd l. Hence, it is equal to unity plus the integer part of $(l/2)$, denoted by $\lfloor l/2\rfloor$. Next, use of the tabulated result [26]

$$\left(\frac{1}{r}\frac{d}{dr}\right)^\nu \frac{H_1^{(2)}}{r} = (-k_0)^\nu\frac{H_{\nu+1}^{(2)}}{r^{\nu+1}} \tag{14}$$

into (12) leads to the expression

$$\frac{\partial^l \phi_0}{\partial \zeta^l} = (-k_0)^l\sum_{m=0}^{\lfloor l/2\rfloor}d_{lm}\frac{\phi_{l-m}}{(k_0\zeta)^m}, \tag{15}$$

and

$$\phi_n = \left(\frac{\zeta}{r}\right)^n H_n^{(2)}(k_0 r). \tag{16}$$

Finally, substituting (2) and (15) into (11) provides

$$S_c(\zeta) = j\pi\Phi_{2i+1,1}, \tag{17}$$

with

$$\Phi_{ln} = \frac{\phi_n}{\alpha^2 k_0\zeta} - \frac{\phi_{n+1}}{\alpha^2} - j\sum_{i=0}^{\infty}\frac{c_{2i}}{\alpha^{2i+1}}\sum_{m=0}^{\lfloor l/2\rfloor}d_{lm}\frac{\phi_{2i-m+n}}{(k_0\zeta)^m}, \tag{18}$$

and hence, the total axial electric field component E_x (1) may be rewritten as

$$E_x = -\frac{\omega\mu_0 I}{4}\left[\phi_0\big|_{\zeta=|z-h|} - (\phi_0 + 2\Phi_{2i+1,1})_{\zeta=z+h}\right].$$ (19)

The non-null magnetic field components may be obtained from E_x by applying Faraday's law, as follows

$$H_y = -\frac{1}{j\omega\mu_0}\frac{\partial E_x}{\partial z} = \frac{I}{2\pi}\left\{\frac{\partial S_d(\zeta)}{\partial\zeta}\bigg|_{\zeta=z-h} - \left[\frac{\partial S_d(\zeta)}{\partial\zeta} - \frac{\partial S_c(\zeta)}{\partial\zeta}\right]_{\zeta=z+h}\right\},$$ (20)

$$H_z = \frac{1}{j\omega\mu_0}\frac{\partial E_x}{\partial y} = -\frac{I}{2\pi}\left\{\frac{\partial S_d(\zeta)}{\partial y}\bigg|_{\zeta=z-h} - \left[\frac{\partial S_d(\zeta)}{\partial y} - \frac{\partial S_c(\zeta)}{\partial y}\right]_{\zeta=z+h}\right\}.$$ (21)

In fact, differentiating (17) with respect to ζ leads to write

$$\frac{\partial S_c(\zeta)}{\partial\zeta} = -j\pi k_0\left(\frac{2\phi_2}{\alpha^2 k_0\zeta} + \Phi_{2i+2,2}\right)$$ (22)

and, as a consequence, (20) is turned into

$$H_y = \frac{jk_0 I}{4}\left[\phi_1\big|_{\zeta=z-h} - \left(\phi_1 + \frac{4\phi_2}{\alpha^2 k_0\zeta} + 2\Phi_{2i+2,2}\right)_{\zeta=z+h}\right].$$ (23)

On the other hand, since it holds

$$\frac{\partial^{l+1}\phi_0}{\partial y\partial\zeta^l} = -k_0\sum_{m=0}^{\lfloor l/2\rfloor}(-1)^m d_{lm}y\zeta^{l-2m}\left(\frac{1}{r}\frac{d}{dr}\right)^{l-m}\frac{H_1^{(2)}}{r}$$

$$= (-k_0)^{l+1}\frac{y}{\zeta}\sum_{m=0}^{\lfloor l/2\rfloor}d_{lm}\frac{\phi_{l-m+1}}{(k_0\zeta)^m}$$ (24)

and, therefore,

$$\frac{\partial S_c(\zeta)}{\partial y} = -\frac{j\pi k_0 y}{\zeta}\Phi_{2i+1,2}(\zeta),$$ (25)

the explicit expression for the H_z-field reads

$$H_z = -\frac{j\pi k_0 y I}{4\pi\zeta}\left[\phi_1\big|_{\zeta=z-h} - (\phi_1 + 2\Phi_{2i+1,2})_{\zeta=z+h}\right].$$ (26)

3. Results

As validation of the theoretical development, expressions (23) and (26) were used to calculate the magnetic field components that an infinite line source carrying 1 A of current produced at a plane 1 m above the interface between air and a clay soil. The source was situated 4 m above the material medium, whose electrical conductivity and dielectric permittivity were taken to be equal to $\sigma_1 = 0.1$ mS/m and $\epsilon_1 = 40\,\epsilon_0$, respectively [27]. At first, the fields were computed against the horizontal distance y from the line axis, assuming that the operating frequency was equal to 1 MHz. Four y-profiles were generated, each one corresponding to a different value for the truncation index L of the outer infinite sum in (18). The results of the computations, shown in Figures 2 and 3, were compared with those arising from numerically evaluating the integral representations for the magnetic field components, namely [28]

$$H_y = \frac{I}{2\pi} \int_0^\infty \left[e^{-u_0|z-h|} - \frac{u_0 - u_1}{u_0 + u_1} e^{-u_0(z+h)} \right] \cos(\lambda y)\, d\lambda, \tag{27}$$

$$H_z = \frac{I}{2\pi} \int_0^\infty \frac{\lambda}{u_0} \left[e^{-u_0|z-h|} + \frac{u_0 - u_1}{u_0 + u_1} e^{-u_0(z+h)} \right] \sin(\lambda y)\, d\lambda. \tag{28}$$

Numerical integration was carried out by applying a Gauss–Kronrod G7-K15 scheme, originating from the combination of a seven-point Gauss rule with a 15-point Kronrod rule. From the analysis of the plotted curves it emerged that increasing the truncation index L improved the accuracy of the result of the computation. In fact, if L grew the curves provided by (23) and (26) approached the outcomes from numerical quadrature, and close agreement was achieved when $L = 9$ in both the situations. Thus, the proposed series-form solution converged to the exact solution. This was confirmed by the curves plotted in Figure 4, which show the relative error of the outcomes from (23) as compared to numerical integration data. As is seen, the error monotonically decreased as L increased. On the other hand, for a fixed value of L the error was not substantially affected by a variation of the distance y as long as the latter was smaller than 50 m. Thereafter, the relative error grew and exhibited an oscillating behavior, pretty similar for all the considered values of the truncation index of the outer sum in (18). However, even with the rapid increase as the distance y increased, for $L = 9$ the error did not exceed 10^{-3} in the considered interval.

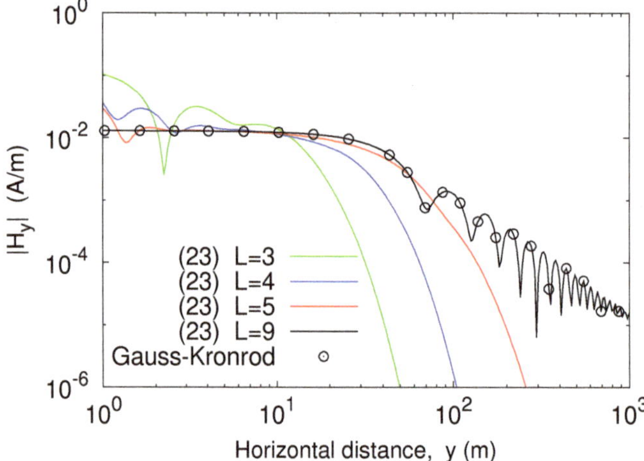

Figure 2. Amplitude of H_y against the horizontal distance from the line source.

The accuracy of the result of the computation also depended on the values of the electromagnetic parameters of the lossy ground and, for instance, better accuracy was observed for larger values of the electrical conductivity σ_1. This aspect is illustrated by Figures 5 and 6, which show, respectively, profiles of $|H_y|$ versus the ratio $\sigma_1 / (\omega \epsilon_1)$, and the relative error resulting from using (23) instead of G7-K15 scheme. Here, it is assumed that the source carried 1 A and operated at 100 kHz, and that $\epsilon_1 = 20\,\epsilon_0$, $h = 5$ m, $y = 10$ m, and $z = 1$ m. As is seen, the sequence of profiles of $|H_y|$ generated by the partial sums in (23) converged to the curve provided by the G7-K15 scheme, and perfect agreement between the analytical and numerical data was again obtained for $L = 9$. Furthermore, convergence was faster in the good-conductor limit ($\sigma_1 \gg \omega \epsilon_1$), where it sufficed to use a sum constituted by seven terms to produce sufficiently accurate results. This does not necessarily imply that the relative error generated by each partial sum in (23) always decreased as the conductivity grew. In fact, as pointed out by Figure 6, the error fluctuated around a decreasing mean value rather than being monotonically decreasing.

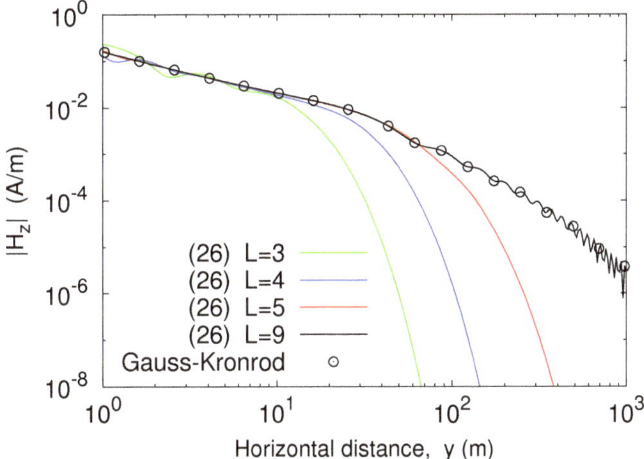

Figure 3. Amplitude of H_z against the horizontal distance from the line source.

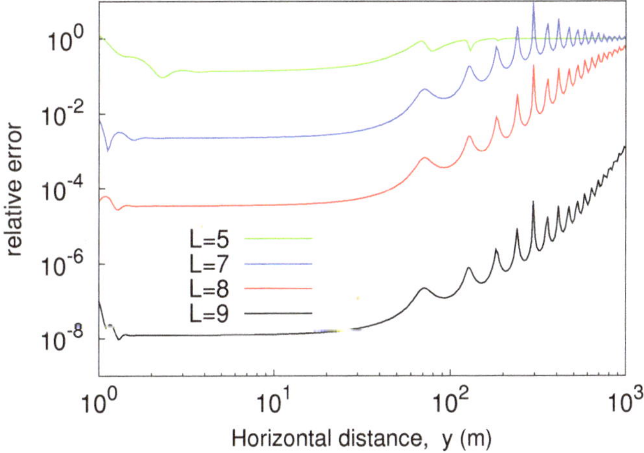

Figure 4. Relative error of (23) as compared to G7-K15 scheme, plotted versus y.

The proposed method made it possible to achieve time savings compared to G7-K15 numerical scheme, while maintaining the same accuracy as the latter. In fact, on a single-core 2.2 GHz processor, the average CPU time taken by (23) to generate the profile of $|H_y|$ associated with $L = 9$ in Figure 5 was equal to 157 ms, while about 9 s were taken by numerical integration of (27) to produce the same outcome. This means that, limited to the considered example, the proposed approach was about 57 times faster than Gaussian integration, and, as a consequence, the numerical complexity of the former was significantly smaller than that exhibited by the latter. For the sake of completeness, the time costs of (23) corresponding to the remaining values of L depicted in Figure 5 are indicated in Table 1.

Table 1. CPU time comparisons for the calculation of H_y.

Approach	Average CPU Time [s]	Speed-Up
G7-K15	8.97	-
(23) with $L = 4$	3.13×10^{-4}	2.87×10^{4}
(23) with $L = 5$	1.26×10^{-3}	7.12×10^{3}
(23) with $L = 7$	3.84×10^{-2}	2.34×10^{2}
(23) with $L = 9$	1.57×10^{-1}	57.1

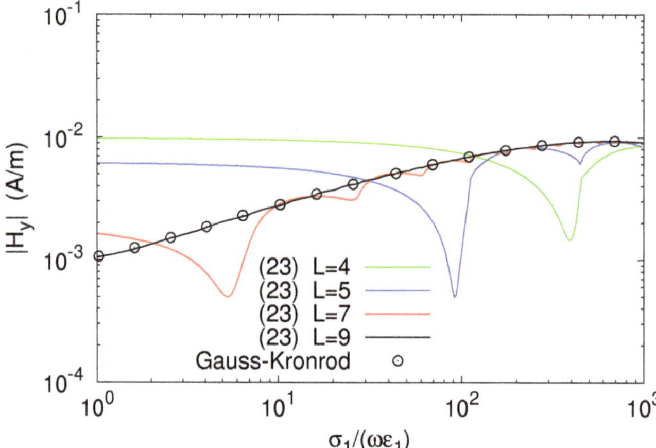

Figure 5. Amplitude of H_y against the normalized conductivity of the ground $\sigma_1/(\omega\epsilon_1)$.

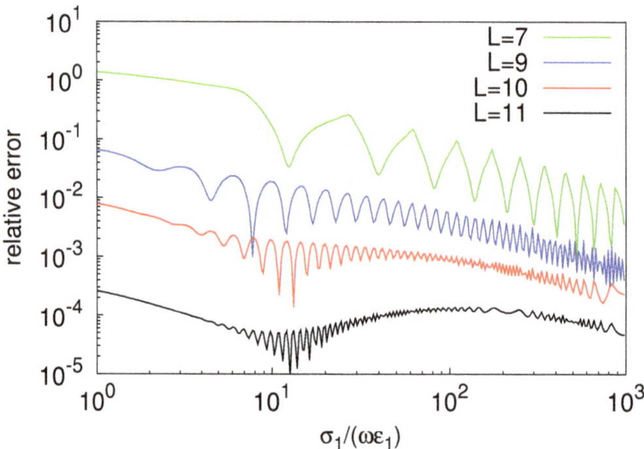

Figure 6. Relative error of (23) as compared to G7-K15 scheme, plotted against $\sigma_1/(\omega\epsilon_1)$.

Finally, one would ask whether the proposed solution may still be used when both h and z approach zero, since it has been obtained starting from expression (9), whose integrals on the right-hand side, strictly speaking, converged only for $\zeta = z + h > 0$. This point is clarified by Figure 7, which shows the amplitude–frequency spectrum of the H_y-field that a unit-current line source generated at a point placed 30 m apart from it, assuming that both the source and the observation point lay on a homogeneous ground with $\sigma_1 = 20$ mS/m and $\epsilon_1 = 10\,\epsilon_0$. As is evident from the analysis of the plotted curves, the trends arising from the sequence of partial sums in (23) still converged to the curve resulting from numerical integration of (27), even if the number of terms of (23) required to approach the numerical data grew as frequency decreased.

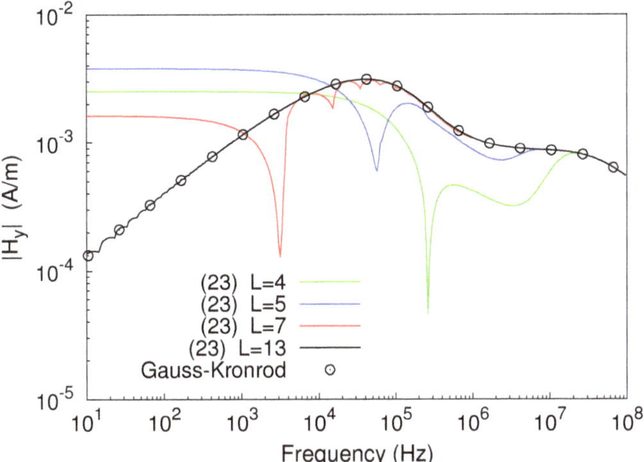

Figure 7. Amplitude-frequency spectrum of H_y, resulting from using both (23) and the G7-K15 scheme.

Hence, the rate of convergence of (23) depended on frequency, and convergence was faster at higher frequencies. This aspect is further investigated in Figure 8, which depicts the relative error of the outcomes of (23) plotted in Figure 7, as compared to the data arising from the G7-K15 rule. The curves of Figure 8 point out that, for a fixed L, the accuracy of the result of the computation significantly worsened when entering the low-frequency range. At the same time, they confirmed that the accuracy could always be enhanced by increasing L. This implies that, all over the considered frequency range, it holds [29]

$$\lim_{L \to \infty} \frac{|H_y^{(L+1)} - H_y|}{|H_y^{(L)} - H_y|^p} = q, \tag{29}$$

where the symbol $\{H_y^{(L)}\}$ denotes the sequence of partial sums that originates from (23), as a result of truncating the infinite sum in (18), and where p and q are the order of convergence (OC) of the sequence and the asymptotic error constant (AEC), respectively. It is easily understood that the parameters p and q depend on the operating frequency, and that the knowledge of their values permits to acquire further information on the rate of convergence. As an example, Table 2 illustrates Lth order estimates of the OC and the AEC for the sequence $\{H_y^{(L)}\}$ corresponding to 1 kHz in Figure 7. The estimates were calculated using the well-known expressions [29]

$$p_L = \frac{\log\left[|H_y^{(L+1)} - H_y^{(L)}|/|H_y^{(L)} - H_y^{(L-1)}|\right]}{\log\left[|H_y^{(L)} - H_y^{(L-1)}|/|H_y^{(L-1)} - H_y^{(L-2)}|\right]}, \tag{30}$$

$$q_L = \frac{|H_y^{(L+1)} - H_y^{(L)}|}{|H_y^{(L)} - H_y^{(L-1)}|^{p_L}} \tag{31}$$

whose limits as $L\to\infty$ approach p and q [29]. As pointed out by Table 2, $p_L\to 1$ with increasing L, and this means that the sequence $\{H_y^{(L)}\}$ converges linearly. Moreover, convergence is accelerated by the small values of the q_L's, which further reduce the remainder $|H_y - H_y^{(L)}|$ at any additional iteration of the sequence.

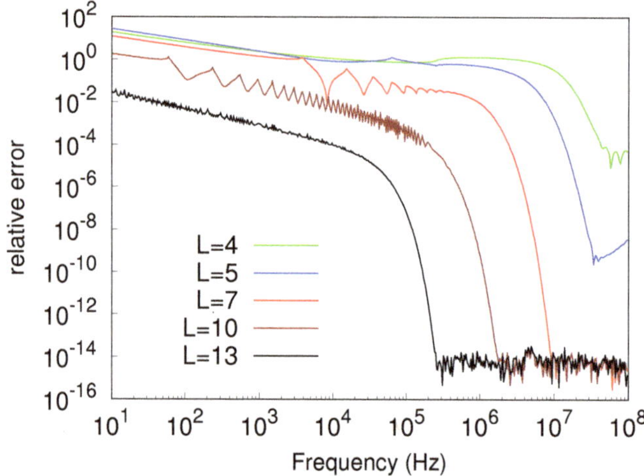

Figure 8. Relative error of (23) as compared to G7-K15 scheme, plotted against frequency.

Table 2. Estimated OC and AEC for the sequence $\{H_y^{(L)}\}$.

L	p_L	q_L
4	0.857	0.674
5	0.871	0.591
7	0.944	0.563
10	0.989	0.495
13	0.996	0.442

4. Conclusions

This paper has presented a rigorous analytical approach for evaluating the time-harmonic EM field components generated in the air space by an infinite current line source located above a homogeneous lossy ground. The approach consists of expanding the non-analytic part of the integrand of the correction term in the integral expression for the axial electric field into a power series of the z-directed propagation coefficient in air. This leads to express the axial electric field as a sum of derivatives of the Sommerfeld Integral describing the direct field, which may be analytically evaluated. As a result, the electric field is given as a sum of cylindrical Hankel functions, with coefficients depending on the position of the field point relative to the line source and its ideal image. Then, explicit expressions for the magnetic field components are also derived by applying Faraday's law. The obtained solution is not subject to simplifying assumptions, and hence is valid even when the high-frequency effects due to the displacement currents in both the air and the soil are no longer negligible. Numerical simulations have been carried out to show that the proposed approach exhibits good accuracy, while taking significantly less computation time than conventional numerical quadrature schemes used to evaluate Sommerfeld-type integrals.

Funding: This research received no external funding.

Conflicts of Interest: The author declares no conflict of interest.

References

1. Alanen, E.; Lindell, I. Image calculation of electromagnetic field from power lines above a dissipative ground. *Arch. Elektrotech.* **1985**, *68*, 259–265. [CrossRef]
2. Rachidi, F.; Tkachenko, S. *Electromagnetic Field Interaction with Transmission Lines: From Classical Theory to HF Radiation Effects*; WIT Press: Southampton, UK, 2008; Volume 5.
3. Spiegel, R.J. Numerical determination of induced currents in humans and baboons exposed to 60-Hz electric fields. *IEEE Trans. Electromagn. Compat.* **1981**, EMC-23, 382–390. [CrossRef]
4. Wait, J.R. Theory of wave propagation along a thin wire parallel to an interface. *Radio Sci.* **1972**, *7*, 675–679. [CrossRef]
5. Wait, J.R.; Spies, K.P. On the image representation of the quasi-static fields of a line current source above the ground. *Can. J. Phys.* **1969**, *47*, 2731–2733. [CrossRef]
6. Olsen, R.G.; Young, J.L.; Chang, D.C. Electromagnetic wave propagation on a thin wire above earth. *IEEE Trans. Antennas Propag.* **2000**, *48*, 1413–1419. [CrossRef]
7. Rachidi, F. A review of field-to-transmission line coupling models with special emphasis to lightning-induced voltages on overhead lines. *IEEE Trans. Electromagn. Compat.* **2012**, *54*, 898–911. [CrossRef]
8. Ametani, A.; Miyamoto, Y.; Baba, Y.; Nagaoka, N. Wave propagation on an overhead multiconductor in a high-frequency region. *IEEE Trans. Electromagn. Compat.* **2014**, *56*, 1638–1648. [CrossRef]
9. Micu, D.D.; Czumbil, L.; Christoforidis, G.C.; Papadopoulos, T. Semi-infinite integral implementation in the development steps of Interfstud electromagnetic interference software. In Proceedings of the IEEE 2012 47th International Universities Power Engineering Conference (UPEC), London, UK, 4–7 September 2012; pp. 1–6.
10. Papakanellos, P.J.; Kaklamani, D.I.; Capsalis, C.N. Analysis of an infinite current source above a semi-infinite lossy ground using fictitious current auxiliary sources in conjunction with complex image theory techniques. *IEEE Trans. Antennas Propag.* **2001**, *49*, 1491–1503. [CrossRef]
11. Wise, W.H. Propagation of HF currents in ground return circuits. *Proc. Inst. Elect. Eng.* **1934**, *22*, 522–527.
12. Kikuchi, H. Wave propagation along an infinite wire above ground at high frequencies. *Proc. Electrotech. J.* **1956**, *2*, 73–78.
13. Dorin, C.; Marilena, U.; Codruta, R. Electromagnetic coupling phenomena of overhead power lines in low and high frequency. In Proceedings of the 2003 IEEE International Symposium on Electromagnetic Compatibility, EMC'03, Istanbul, Turkey, 11–16 May 2003; Volume 2, pp. 1178–1181.
14. Degauque, P.; Laly, P.; Degardin, V.; Lienard, M. Power line communication and compromising radiated emission. In Proceedings of the IEEE SoftCOM 2010, 18th International Conference on Software, Telecommunications and Computer Networks, Split, Croatia, 23–25 September 2010; pp. 88–91.
15. Pagani, P.; Ney, M.; Zeddam, A. Application of Time Reversal to Power Line Communications for the Mitigation of Electromagnetic Radiation. In *Electromagnetic Time Reversal: Application to EMC and Power Systems*; Rachidi, F., Rubinstein, M., Paolone, M., Eds.; Wiley Online Library: New York, NY, USA, 2017; Chapter 5, pp. 169–187.
16. Sunde, E.D. *Earth Conduction Effects in Transmission Systems*; Dover: New York, NY, USA, 1968.
17. Pistol'kors, A. On the theory of a wire parallel to the plane interface between two media. *Radiotek* **1953**, *8*, 8–18.
18. Kuester, E.F.; Chang, D.C.; Olsen, R.G. Modal theory of long horizontal wire structures above the earth—Part I: Excitation. *Radio Sci.* **1978**, *13*, 605–613. [CrossRef]
19. Judkins, R.; Nordell, D. Discussion of Electromagnetic Effects of Overhead Transmission Lines Practical Problems, Safeguards and Methods of Calculation. *IEEE Trans. Power Apparatus Syst.* **1974**, PAS-93, 892–902.
20. Kostenko, M. Mutual impedance of earth-return overhead lines taking into account the skin-effect. *Elektritchestvo* **1955**, *10*, 29–34.
21. Chang, D.C.; Olsen, R.G. Excitation of an infinite antenna above a dissipative earth. *Radio Sci.* **1975**, *10*, 823–831. [CrossRef]
22. Olsen, R.G.; Kuester, E.F.; Chang, D.C. Modal theory of long horizontal wire structures above the earth—Part II: Properties of discrete modes. *Radio Sci.* **1978**, *13*, 615–623. [CrossRef]
23. Déri, Á.; Tevan, G. Mathematical verification of Dubanton's simplified calculation of overhead transmission line parameters and its physical interpretation. *Arch. Elektrotech.* **1981**, *63*, 191–198. [CrossRef]

24. Tevan, G.; Deri, A. Some remarks about the accurate evaluation of the Carson integral for mutual impedances of lines with earth return. *Arch. Elektrotech.* **1984**, *67*, 83–90. [CrossRef]
25. Mohsen, A.; Shafai, L. On the image representation of the fields of a line current source above finitely conducting earth. *Can. J. Phys.* **1981**, *59*, 117–121. [CrossRef]
26. Abramowitz, M.; Stegun, I.A. *Handbook of Mathematical Functions: With Formulas, Graphs, and Mathematical Tables*; Courier Corporation: Massachusetts, MA, USA, 1964; Volume 55.
27. Palacky, G.J. Resistivity characteristics of geologic targets. In *Electromagnetic Methods in Applied Geophysics*; Nabighian, M.N., Ed.; Society of Exploration Geophysicists: Tulsa, OK, USA, 1988; Chapter 3, Volume 1, pp. 52–129. [CrossRef]
28. Ward, S.H.; Hohmann, G.W. Electromagnetic theory for geophysical applications. In *Electromagnetic Methods in Applied Geophysics*; Nabighian, M.N., Ed.; Society of Exploration Geophysicists: Tulsa, OK, USA, 1988; Chapter 4, Volume 1, pp. 130–311. [CrossRef]
29. Householder, A.S. *The Numerical Treatment of a Single Nonlinear Equation*; McGraw-Hill: New York, NY, USA, 1970.

Publisher's Note: MDPI stays neutral with regard to jurisdictional claims in published maps and institutional affiliations.

 *electronics*

Article

Cylindrical Waveguide on Ferrite Substrate Controlled by Externally Applied Magnetic Field

Hedi Sakli [1,2]

1 MACS Laboratory: Modeling, Analysis and Control of Systems RL16ES22, National Engineering School of Gabes, University of Gabes, Gabes 6029, Tunisia; hedi.s@eitaconsulting.fr
2 EITA Consulting, 5 Rue du Chant des Oiseaux, 78360 Montesson, France

Abstract: This paper presents an extension of the formulation of wave propagation in transverse electric (TE) and transverse magnetic (TM) modes for the case of metallic cylindrical waveguides filled with longitudinally magnetized ferrite. The higher order modes were exploited. We externally controlled the cut-off frequency through the application of DC magnetic fields. The numerical results of dispersion diagrams for TE and TM modes were obtained and analyzed. We analyzed a waveguide antenna filled with partially magnetized ferrite using the mode matching (MM) technique based on the TE and TM modes. By using modal analysis, our approach considerably reduced the computation time compared to HFSS. Ferrites are important for various industrial applications, such as circulators, isolators, antennas and filters.

Keywords: anisotropic materials; antenna; cylindrical waveguides; ferrites; propagation

Citation: Sakli, H. Cylindrical Waveguide on Ferrite Substrate Controlled by Externally Applied Magnetic Field. *Electronics* **2021**, *10*, 474. https://doi.org/10.3390/electronics10040474

Academic Editor: Giulio Antonini
Received: 20 January 2021
Accepted: 13 February 2021
Published: 17 February 2021

Publisher's Note: MDPI stays neutral with regard to jurisdictional claims in published maps and institutional affiliations.

1. Introduction

Recently, many researchers have been interested in guiding devices that use ferrite in some frequency range for their potential applications in microwave circuits. However, there is a lack of research into the dispersion of ferrite cylindrical waveguides. Among the essential research pertaining to this topic, we can cite the work in [1–12]. Guided modes in waveguides consisting of anisotropic media [4,13–15] have been studied in the literature.

In this paper, we present an extension of the transverse electric (TE) and transverse magnetic (TM) modes to cylindrical waveguides filled with lossless longitudinally magnetized ferrite (LMF) which takes account of the spatial distribution of the permeability of the medium that is applied to the transverse fields. We exploited the propagation modes in this structure. We show how the dispersion diagrams were obtained and we discuss the effects of anisotropic parameters on dispersion characteristics and cutoff frequencies. We also show how the numerical results for the TE and TM modes were obtained. These modes were used in the numerical method applied to the design of our antenna. Our simulations of a cylindrical metallic waveguide antenna filled with partially magnetized ferrite using mode matching (MM) [16,17] were in good agreement with those obtained with HFSS. However, our method was noticeably faster.

Our objective was to analyze the physical discontinuities in a cylindrical metallic waveguide filled with partially magnetized ferrite using the mode matching technique based on the TE and TM modes. An antenna formed by this type of waveguide is presented and analyzed. The magnetization of the ferrite using an external control can be achieved by enveloping the waveguide with Helmholtz coils connected to a variable voltage generator. Each voltage corresponds to a constant magnetic field applied to the ferrite longitudinally which creates the magnetization of this medium. As a result, the magnetic properties of the ferrite change. The permeability of the ferrite becomes tensorial. Our antenna was tuned to the desired operating frequency for the ferrite magnetization circuit by adjusting the voltage applied externally, which in turn affected the variation of the static magnetic field applied to the ferrite. The operating frequency range could vary. Our interest in this study was in the tuning of the operating frequency of the antenna.

This formulation is a useful tool for microwave engineers. This type of material is extensively applied by information technology industries, particularly in microwaves and RF devices, such as patch antennas, waveguide antennas, resonators, circulators, insulators, phase converters and filters.

2. Formulation

For lossless longitudinally magnetized ferrite, as is shown in Figure 1, the Maxwell equations can be written as:

$$\vec{\nabla} \times \vec{E} = -j\omega\overline{\mu}_f\,\vec{H} \tag{1}$$

$$\vec{\nabla} \times \vec{H} = j\omega\varepsilon_f\,\vec{E} \tag{2}$$

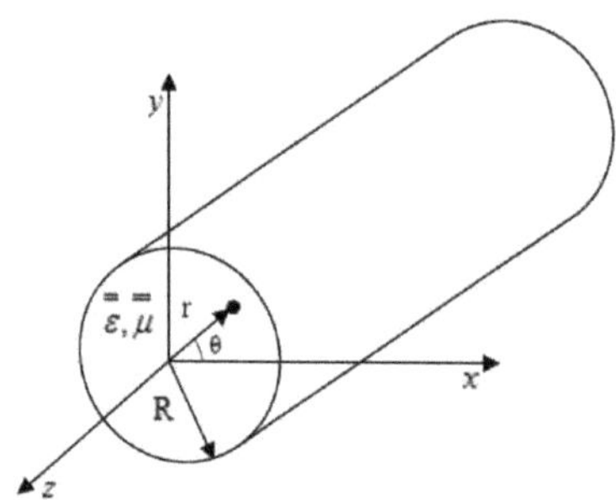

Figure 1. Geometry of cylindrical waveguide filled with longitudinally magnetized ferrite (LMF).

At the microwave's frequencies, the ferrites are characterized by a tensor permeability that represents their induced anisotropy under a magnetic field. The permeability of the LMF is tensorial and can be written in a system of cylindrical coordinates, as is given by D. Polder [8], with

$$\overline{\mu}_f = \mu_0\,\overline{\mu}_{rf} = \mu_0 \begin{pmatrix} \mu & -j\kappa & 0 \\ j\kappa & \mu & 0 \\ 0 & 0 & \mu_{rz} \end{pmatrix} = \mu_0 \begin{pmatrix} \mu_{rT} & 0 \\ 0 & \mu_{rz} \end{pmatrix} \tag{3}$$

where μ, κ, μ_{rz} and ε_f are real quantities.

For a partial magnetization of ferrite, J. J. Green et al. [5] and E. Schloemann [9] give the empirical expressions of μ, κ, μ_{rz} [11]:

$$\mu = \mu_d + (1 - \mu_d)\left(\frac{4\pi M}{4\pi M_S}\right)^{3/2} \tag{4}$$

$$\kappa = \frac{\gamma(4\pi M)}{\omega} \tag{5}$$

$$\mu_{rz} = \mu_d^{\left(1 - \left(\frac{4\pi M}{4\pi M_S}\right)^{5/2}\right)} \tag{6}$$

where

$$\mu_d = \frac{1}{3}\left\{1 + 2\sqrt{1 - \left(\frac{\gamma(4\pi M_S)}{\omega}\right)^2}\right\} \tag{7}$$

where ω is the work pulsation, γ is the gyromagnetic constant, $4\pi M_S$ is the magnetization at saturation and $4\pi M$ is the magnetization which is lower than saturation. When the magnetization is equal to zero, $\kappa = 0$ and $\mu = \mu_{rz} = 1$. The ferrite then becomes an isotropic dielectric.

Let us consider a cylindrical waveguide of radius R completely filled with LMF without losses, as represented in the Figure 1. The walls of the guide are perfectly electric conductive. In this study, we rigorously examined the TE and TM modes in a metallic cylindrical waveguide completely full of longitudinally magnetized ferrite.

2.1. Transverse Electric Modes

By considering the propagation in the Oz direction and manipulating Equations (1) and (2), we obtain the expressions of the transverse electromagnetic fields according to the longitudinal fields in the TE modes.

$$E_r^{(h)} = \frac{-1}{K_c^2}\left(A_1\frac{\partial H_z}{\partial r} + jA_2\frac{1}{r}\frac{\partial H_z}{\partial \theta}\right) \tag{8}$$

$$E_\theta^{(h)} = \frac{1}{K_c^2}\left(jA_2\frac{\partial H_z}{\partial r} - A_1\frac{1}{r}\frac{\partial H_z}{\partial \theta}\right) \tag{9}$$

$$H_r^{(h)} = \frac{1}{K_c^2}\left(-jk_z K_{c\mu}^2\frac{\partial H_z}{\partial r} + Fk_z\frac{1}{r}\frac{\partial E_z}{\partial \theta}\right) \tag{10}$$

$$H_\theta^{(h)} = \frac{-1}{K_c^2}\left(Fk_z\frac{\partial H_z}{\partial r} + jk_z K_{c\mu}^2\frac{1}{r}\frac{\partial H_z}{\partial \theta}\right) \tag{11}$$

with

$$K_{c\mu}^2 = k_0^2 \varepsilon_{rf}\mu - k_z^2 \tag{12}$$

$$F = k_0^2 \varepsilon_{rf}\kappa \tag{13}$$

$$K_c^2 = k_{c\mu}^4 - F^2 \tag{14}$$

$$k_0^2 = \omega^2 \varepsilon_0 \mu_0 \tag{15}$$

$$A_1 = \frac{Fk_z^2}{\omega\varepsilon_0\varepsilon_{rf}} \tag{16}$$

$$A_2 = \frac{k_{c\mu}^4 - F^2 + k_z^2 k_{c\mu}^2}{\omega\varepsilon_0\varepsilon_{rf}} \tag{17}$$

From Equation (1), the differential equation for z-component can be obtained as follows

$$\frac{\partial^2 H_z}{\partial r^2} + \frac{1}{r}\frac{\partial H_z}{\partial r} + \frac{1}{r^2}\frac{\partial^2 H_z}{\partial \theta^2} + \left(K_{cf}^{(h)}\right)^2 H_z = 0 \tag{18}$$

with

$$\left(K_{cf}^{(h)}\right)^2 = \frac{\omega\mu_0\mu_{rz}K_c^2}{A_2} \tag{19}$$

The resolution of the differential Equation (18), due to the separation of the two variables, requires the expression of H_z for the TE_{mn} modes in the metallic cylindrical waveguide fully filled with LMF. The expression of the longitudinal magnetic field can be written as follows

$$H_z^{(h)} = H_0 sin(n\theta)J_n\left(K_{cf}^{(h)}r\right)e^{j(\omega t - k_z^{(h)}z)} \tag{20}$$

J_n is the Bessel function of the first kind of order n $(n = 0, 1, 2, 3, \ldots)$.
Equations (8)–(11) become

$$E_r^{(h)} = \frac{H_0}{K_c^2}\left\{-A_1 K_{cf}^{(h)} sin(n\theta)J_n'\left(K_{cf}^{(h)}r\right) - jA_2\frac{n}{r}cos(n\theta)J_n\left(K_{cf}^{(h)}r\right)\right\} \tag{21}$$

$$E_\theta^{(h)} = \frac{H_0}{K_c^2}\left\{ jA_2 K_{cf}^{(h)} \sin(n\theta) J_n'\left(K_{cf}^{(h)} r\right) - A_1 \frac{n}{r}\cos(n\theta) J_n\left(K_{cf}^{(h)} r\right)\right\} \tag{22}$$

$$H_r^{(h)} = \frac{H_0}{K_c^2}\left\{ -jk_z^{(h)} K_{c\mu}^2 K_{cf}^{(h)} \sin(n\theta) J_n'\left(K_{cf}^{(h)} r\right) + Fk_z^{(h)}\frac{n}{r}\cos(n\theta) J_n\left(K_{cf}^{(h)} r\right)\right\} \tag{23}$$

$$H_\theta^{(h)} = \frac{H_0}{K_c^2}\left\{ -Fk_z^{(h)} K_{cf}^{(h)} \sin(n\theta) J_n'\left(K_{cf}^{(h)} r\right) - jk_z^{(h)} K_{c\mu}^2 \frac{n}{r}\cos(n\theta) J_n\left(K_{cf}^{(h)} r\right)\right\} \tag{24}$$

J_n' is the derivative of the Bessel function of the first kind of order n ($n = 0, 1, 2, 3, \dots$).
The boundary conditions give

$$J_n'\left(u_{nm}'\right) = 0 \tag{25}$$

with

$$u_{nm}' = K_{cf}^{(h)} R \tag{26}$$

In Equation (26), u_{nm}' represents the mth zero ($m = 1, 2, 3, \dots$) of the derivative of the Bessel function J_n' of the first kind of order n. The determination of constant H_0 is done by normalizing the power P^{TE} that crosses the cross-section of the guide, which is in our case a disc of radius R.

$$P^{TE} = \int_0^R \int_0^{2\pi}\left(E_r^{(h)} H_\theta^{*(h)} - E_\theta^{(h)} H_r^{*(h)}\right) r\, dr\, d\theta = 1 \tag{27}$$

$*$ indicates the complex conjugate.
Equation (27) gives

$$H_0 = \frac{K_c^2 K_{cf}}{\sqrt{k_z\left(A_2 K_{c\mu}^2 + A_1 F\right)}} N_{nm}^{(h)} \tag{28}$$

with

$$N_{nm}^{(h)} = \frac{1}{\sqrt{\frac{\sigma_n}{2}}\left((u_{nm}')^2 - n^2\right)^{1/2} J_n(u_{nm}')} \tag{29}$$

$$\sigma_n = \begin{cases} 2\pi & if\ n = 0 \\ \pi & if\ n > 0 \end{cases} \tag{30}$$

From Equation (26), we obtain the propagation equation

$$k_z^4 + \left(-2\omega^2 \varepsilon_0 \mu_0 \varepsilon_{rf}\mu + \frac{\mu}{\mu_{rz}}\left(\frac{u_{nm}'}{R}\right)^2\right) k_z^2$$
$$+ \omega^2 \varepsilon_0 \mu_0 \varepsilon_{rf}\left(\mu^2 - \kappa^2\right)\left[\omega^2 \varepsilon_0 \mu_0 \varepsilon_{rf} - \frac{1}{\mu_{rz}}\left(\frac{u_{nm}'}{R}\right)^2\right] = 0 \tag{31}$$

To find cut-off frequencies, the propagation constant should be equated to zero in Equation (31). The cutoff frequency in the TE mode is written

$$f_{c.nm}^{(TE)} = \frac{c}{2\pi}\frac{1}{\sqrt{\varepsilon_{rf}\mu_{rz}}}\left(\frac{u_{nm}'}{R}\right) \tag{32}$$

If the magnetization $4\pi M$ decreases, μ_{rz} increases. Consequently, the cutoff frequency $f_{c.nm}^{(TE)}$ decreases. We can externally control the cut-off frequency through the application of DC magnetic fields. The magnetization of the ferrite using an external control can be achieved by enveloping the waveguide with Helmholtz coils connected to a variable voltage generator.

2.2. Transverse Magnetic (TM) Modes

By manipulating Equations (1) and (2), we can obtain the expressions of the transverse electromagnetic fields according to the longitudinal fields in the TM modes.

$$E_r^{(e)} = \frac{k_z}{K_c^2}\left(-jK_{c\mu}^2\frac{\partial E_z}{\partial r} + F\frac{1}{r}\frac{\partial E_z}{\partial \theta}\right) \tag{33}$$

$$E_\theta^{(e)} = \frac{-k_z}{K_c^2}\left(F\frac{\partial E_z}{\partial r} + jK_{c\mu}^2\frac{1}{r}\frac{\partial E_z}{\partial \theta}\right) \tag{34}$$

$$H_r^{(e)} = \frac{\omega\varepsilon_0\varepsilon_{rf}}{K_c^2}\left(F\frac{\partial E_z}{\partial r} + jK_{c\mu}^2\frac{1}{r}\frac{\partial E_z}{\partial \theta}\right) \tag{35}$$

$$H_\theta^{(e)} = \frac{\omega\varepsilon_0\varepsilon_{rf}}{K_c^2}\left(-jK_{c\mu}^2\frac{\partial E_z}{\partial r} + F\frac{1}{r}\frac{\partial E_z}{\partial \theta}\right) \tag{36}$$

From Equation (1), the differential equation for the z component can be obtained as follows

$$\frac{\partial^2 E_z}{\partial r^2} + \frac{1}{r}\frac{\partial E_z}{\partial r} + \frac{1}{r^2}\frac{\partial^2 E_z}{\partial \theta^2} + \left(K_{cf}^{(e)}\right)^2 E_z = 0 \tag{37}$$

with

$$K_{cf}^{(e)} = \frac{K_c}{K_{c\mu}} \tag{38}$$

The resolution of the differential Equation (37), due to the separation of the two variables, requires the expression of E_z for the TM_{nm} modes in the metallic cylindrical waveguide fully filled with LMF. The expression of the longitudinal electric field can be written as follows

$$E_z^{(e)} = E_0 cos(n\theta) J_n\left(K_{cf}^{(e)} r\right) e^{j(\omega t - k_z^{(e)} z)} \tag{39}$$

Equations (33)–(36) become

$$E_r^{(e)} = \frac{-E_0 k_z}{K_c^2}\left\{jK_{c\mu}^2 K_{cf}^{(e)} cos(n\theta) J_n'\left(K_{cf}^{(e)} r\right) + F\frac{n}{r}sin(n\theta) J_n\left(K_{cf}^{(e)} r\right)\right\} \tag{40}$$

$$E_\theta^{(e)} = \frac{E_0 k_z}{K_c^2}\left\{-FK_{cf}^{(e)} cos(n\theta) J_n'\left(K_{cf}^{(e)} r\right) + jK_{c\mu}^{(2)}\frac{n}{r}sin(n\theta) J_n\left(K_{cf}^{(e)} r\right)\right\} \tag{41}$$

$$H_r^{(e)} = \frac{\omega\varepsilon_0\varepsilon_{rf} E_0}{K_c^2}\left\{FK_{cf}^{(e)} cos(n\theta) J_n'\left(K_{cf}^{(e)} r\right) - jK_{c\mu}^2\frac{n}{r}sin(n\theta) J_n\left(K_{cf}^{(e)} r\right)\right\} \tag{42}$$

$$H_\theta^{(e)} = \frac{-\omega\varepsilon_0\varepsilon_{rf} E_0}{K_c^2}\left\{jK_{c\mu}^2 K_{cf}^{(e)} cos(n\theta) J_n'\left(K_{cf}^{(e)} r\right) + F\frac{n}{r}sin(n\theta) J_n\left(K_{cf}^{(e)} r\right)\right\} \tag{43}$$

The boundary conditions give the following equation

$$J_n(u_{nm}) = 0 \tag{44}$$

with

$$u_{nm} = K_{cf}^{(e)}.R \tag{45}$$

In Equation (45), u_{nm} represents the mth zero ($m = 1, 2, 3, \dots$) of the Bessel function J_n of the first kind of order n. The determination of constant E_0 is done by normalizing the power P^{TM} that crosses the cross-section of the guide, which is in our case a disc of radius R.

$$P^{TM} = \int_0^R \int_0^{2\pi} \left(E_r^{(e)} H_\theta^{*(e)} - E_\theta^{(e)} H_r^{*(e)}\right) r dr d\theta = 1 \tag{46}$$

Equation (46) gives

$$E_0 = \frac{1}{\sqrt{\omega \varepsilon_0 \varepsilon_{rf} k_z}} \frac{K_c^2}{\sqrt{K_{c\mu}^4 + F^2}} N_{nm}^{(e)}$$ (47)

with

$$N_{nm}^{(e)} = \frac{1}{u_{nm} \cdot J_n'(u_{nm}) \cdot \sqrt{\frac{\sigma_n}{2}}}$$ (48)

Finally, the propagation constant in the TM mode is given by

$$k_z^4 + \left(-2\omega^2 \varepsilon_0 \mu_0 \varepsilon_{rf} \mu + \left(\tfrac{u_{nm}}{R}\right)^2 \right) k_z^2 \cos^{-1} \theta$$
$$+ \omega^2 \varepsilon_0 \mu_0 \varepsilon_{rf} \left[\omega^2 \varepsilon_0 \mu_0 \varepsilon_{rf} (\mu^2 - \kappa^2) - \mu \left(\tfrac{u_{nm}}{R}\right)^2 \right] = 0.$$ (49)

Obviously, the cutoff frequency is written

$$f_{c.nm}^{(TM)} = \frac{c}{2\pi} \frac{1}{\sqrt{\mu_{reff} \varepsilon_{rf}}} \cdot \left(\frac{u_{nm}}{R} \right)$$ (50)

with

$$\mu_{reff} = \frac{\mu^2 - \kappa^2}{\mu}$$ (51)

μ_{reff} is the effective permeability.

If the magnetization $4\pi M$ decreases, μ_{reff} increases. Consequently, the cutoff frequency $f_{c.nm}^{(TM)}$ in the TE modes decreases. We can externally control the cut-off frequency through the application of DC magnetic fields.

3. Analysis of Uni-Axial Discontinuities in the Cylindrical Waveguides

In this section, we describe how the use of MM can be extended to characterize uni-axial discontinuities between metallic cylindrical waveguides filled with the studied medium (LMF). The discontinuities were considered without losses. This method was based on the modal development of the transverse electromagnetic fields.

Figure 2 shows a junction between two cylindrical waveguides filled with two different media with the same cross-sections, where a^i and b^i are the incident and the reflected waves, respectively.

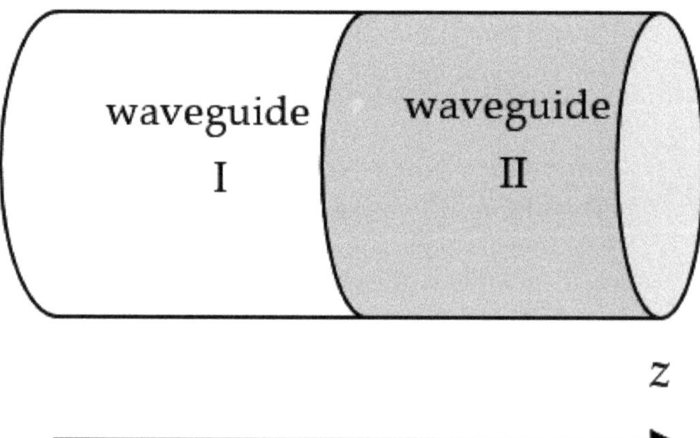

Figure 2. A junction between two cylindrical waveguides filled with two different media with the same cross-sections.

The transverse electric and magnetic fields (E_T, H_T) in the waveguides can be written in the modal bases as follows [16,17].

$$E_T = \sum_{m=1}^{\infty} A_m^i \left(a_m^i + b_m^i \right) e_m^i \tag{52}$$

$$H_T = \sum_{m=1}^{\infty} B_m^i \left(a_m^i - b_m^i \right) h_m^i \tag{53}$$

where E_T and H_T are the transverse electric and magnetic fields (the sub-index T refers to the components in the transverse plane) and A_m^i and B_m^i are complex coefficients which are determined by normalizing the power flow down the cylindrical guides ($i = I, II$ and m is the index of the mode). e_m^i and h_m^i represent the mth electric and magnetic modal eigenfunction in the guide i, respectively.

At the junction, the continuity of the fields makes it possible to write the following equations:

$$E_T^I = E_T^{II} \tag{54}$$

$$H_T^I = H_T^{II} \tag{55}$$

By inserting Equations (52) and (53) into Equations (54) and (55), we obtain:

$$\sum_{m=1}^{N_1} A_m^I \left(a_m^I + b_m^I \right) e_m^I = \sum_{p=1}^{N_2} A_p^{II} \left(a_p^{II} + b_p^{II} \right) e_p^{II} \tag{56}$$

$$\sum_{m=1}^{N_1} B_m^I \left(a_m^I - b_m^I \right) h_m^I = \sum_{p=1}^{N_2} B_p^{II} \left(-a_p^{II} + b_p^{II} \right) h_p^{II} \tag{57}$$

N_1 and N_2 are the numbers of considered modes in guides 1 and 2, respectively. By applying Galerkin's method, Equations (56) and (57) lead to the following systems:

$$\sum_{m=1}^{N_1} A_m^I \left(a_m^I + b_m^I \right) \left\langle e_m^I \middle| e_p^{II} \right\rangle = A_p^{II} \left(a_p^{II} + b_p^{II} \right) \tag{58}$$

$$B_m^I \left(a_m^I - b_m^I \right) = \sum_{p=1}^{N_2} B_p^{II} \left(-a_p^{II} + b_p^{II} \right) \left\langle h_p^{II} \middle| h_m^I \right\rangle \tag{59}$$

The inner product can be defined as:

$$\left\langle e_m \middle| e_p \right\rangle = \int_S e_m^* e_p \, dS \tag{60}$$

Equations (58) and (59) give:

$$-a_p^{II} + \sum_{m=1}^{N_1} \frac{A_m^I}{A_p^{II}} a_m^I \left\langle e_m^I \middle| e_p^{II} \right\rangle = b_p^{II} - \sum_{m=1}^{N_1} \frac{A_m^I}{A_p^{II}} b_m^I \left\langle e_m^I \middle| e_p^{II} \right\rangle \tag{61}$$

$$a_m^I + \sum_{p=1}^{N_2} \frac{B_p^{II}}{B_m^I} a_p^{II} \left\langle h_p^{II} \middle| h_m^I \right\rangle = b_m^I + \sum_{p=1}^{N_2} \frac{B_p^{II}}{B_m^I} b_p^{II} \left\langle h_p^{II} \middle| h_m^I \right\rangle \tag{62}$$

which can be written in matrix form:

$$\begin{bmatrix} U & M_1 \\ M_2 & -U \end{bmatrix} \begin{bmatrix} a_1^I \\ \cdot \\ a_{N_1}^I \\ a_1^{II} \\ \cdot \\ a_{N_2}^{II} \end{bmatrix} = \begin{bmatrix} U & M_1 \\ -M_2 & U \end{bmatrix} \begin{bmatrix} b_1^I \\ \cdot \\ b_{N_1}^I \\ b_1^{II} \\ \cdot \\ b_{N_2}^{II} \end{bmatrix} \tag{63}$$

where U is the identity matrix. M_1 and M_2 (of dimensions $(N_1 \times N_2)$ and $(N_2 \times N_1)$ respectively) are matrixes of the following general terms:

$$M_{1ij} = \frac{B_j^{II}}{B_i^I} \left\langle h_j^{II} \middle| h_i^I \right\rangle \tag{64}$$

$$M_{2ij} = \frac{A_i^I}{A_j^{II}} \left\langle e_i^I \middle| e_j^{II} \right\rangle \tag{65}$$

The scattering matrix of the discontinuity is:

$$S = \begin{bmatrix} U & M_1 \\ -M_2 & U \end{bmatrix}^{-1} \begin{bmatrix} U & M_1 \\ M_2 & -U \end{bmatrix} \tag{66}$$

of dimensions $((N_1 + N_2) \times (N_1 + N_2))$.

In the numerical calculations we have to invert a complex matrix of dimensions equal to the sum of the modes taken into account on each side of the discontinuity.

For a structure with multiple uni-axial discontinuities in cascade, the total matrix can be obtained by separating the chaining of S matrixes of discontinuities with waveguides of lengths equal to the distances between the discontinuities.

Figure 3 represents a double discontinuity. From the incident and reflected waves, we can write the following equations

$$\left(\begin{bmatrix} b_1^I \\ b_2^I \end{bmatrix} \right) = S^I \left(\begin{bmatrix} a_1^I \\ a_2^I \end{bmatrix} \right) \tag{67}$$

$$\left(\begin{bmatrix} b_1^{II} \\ b_2^{II} \end{bmatrix} \right) = S^{II} \left(\begin{bmatrix} a_1^{II} \\ a_2^{II} \end{bmatrix} \right) \tag{68}$$

$$\begin{bmatrix} b_1^I \end{bmatrix} = S_{11}^I \begin{bmatrix} a_1^I \end{bmatrix} + S_{12}^I \begin{bmatrix} a_2^I \end{bmatrix} \tag{69}$$

$$\begin{bmatrix} b_2^I \end{bmatrix} = S_{21}^I \begin{bmatrix} a_1^I \end{bmatrix} + S_{22}^I \begin{bmatrix} a_2^I \end{bmatrix} \tag{70}$$

$$\begin{bmatrix} b_1^{II} \end{bmatrix} = S_{11}^{II} \begin{bmatrix} a_1^{II} \end{bmatrix} + S_{12}^{II} \begin{bmatrix} a_2^{II} \end{bmatrix} \tag{71}$$

$$\begin{bmatrix} b_2^{II} \end{bmatrix} = S_{21}^{II} \begin{bmatrix} a_1^{II} \end{bmatrix} + S_{22}^{II} \begin{bmatrix} a_2^{II} \end{bmatrix} \tag{72}$$

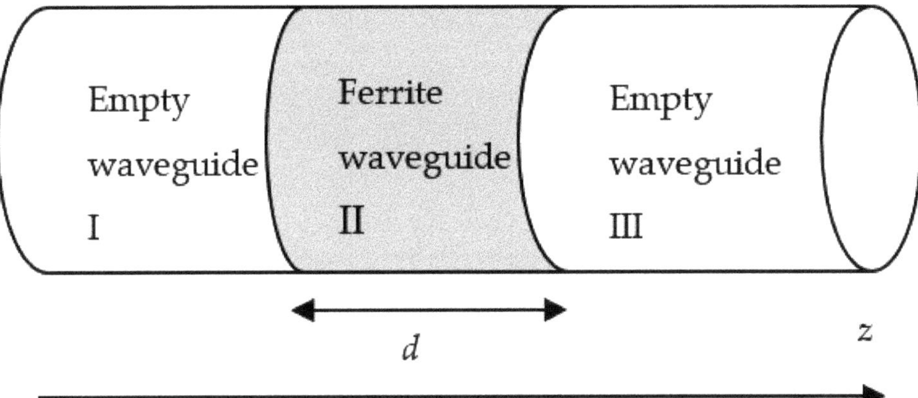

Figure 3. Cylindrical waveguide antenna filled with partially magnetized ferrite of width d with the same cross-section as the other waveguides.

There are

$$\left[a_2^I\right] = D\left[b_2^{II}\right] \tag{73}$$

$$\left[a_2^{II}\right] = D\left[b_2^I\right] \tag{74}$$

with

$$D = \begin{bmatrix} e^{-\gamma_1.d} & 0 & 0 \\ 0 & \dots & 0 \\ 0 & 0 & e^{-\gamma_N.d} \end{bmatrix} \tag{75}$$

γ_i is the propagation constant of the ith mode of the central guide and N is the number of modes in the same guide.

Using Equations (73) and (74), we get

$$\left[b_1^I\right] = S_{11}^I\left[a_1^I\right] + S_{12}^I D\left[b_2^{II}\right] \tag{76}$$

$$\left[b_1^{II}\right] = S_{11}^{II}\left[a_1^{II}\right] + S_{12}^{II} D\left[b_2^I\right] \tag{77}$$

$$\left[b_2^I\right] = S_{21}^I\left[a_1^I\right] + S_{22}^I D\left[b_2^{II}\right] \tag{78}$$

$$\left[b_2^{II}\right] = S_{21}^{II}\left[a_1^{II}\right] + S_{22}^{II} D\left[b_2^I\right] \tag{79}$$

Equation (79), in using Equation (78), becomes

$$\left[b_2^{II}\right] = S_{21}^{II}\left[a_1^{II}\right] + S_{22}^{II} D S_{21}^I\left[a_1^I\right] + S_{22}^{II} D S_{22}^I D\left[b_2^{II}\right] \tag{80}$$

We put

$$E = \left[U - S_{22}^{II} D S_{22}^I D\right]^{-1} \tag{81}$$

U is the identity matrix. As a result

$$\left[b_2^{II}\right] = E S_{22}^{II} D S_{21}^I\left[a_1^I\right] + E S_{21}^{II}\left[a_1^{II}\right] \tag{82}$$

$$\left[b_1^I\right] = \left[S_{11}^I + S_{12}^I D E S_{22}^{II} D S_{21}^I\right]\left[a_1^I\right] \\ + \left[S_{12}^I D E S_{21}^{II}\right]\left[a_1^{II}\right] \tag{83}$$

$$[b_1^{II}] = S_{12}^I D \left[U + S_{22}^I DES_{22}^{II} D \right] S_{21}^I [a_1^I]$$
$$+ \left[S_{11}^{II} + S_{12}^{II} DS_{22}^I DES_{21}^{II} \right] [a_1^{II}].$$
(84)

The matrix S of the double discontinuity is given by

$$S = \begin{bmatrix} S_{11}^I + S_{12}^I DES_{22}^{II} DS_{21}^I & S_{12}^I DES_{21}^{II} \\ S_{12}^I D \left[U + S_{22}^I DES_{22}^{II} D \right] S_{21}^I & S_{11}^{II} + S_{12}^{II} DS_{22}^I DES_{21}^{II} \end{bmatrix}$$
(85)

Thus, the matrix S of several discontinuities in cascade can be determined from Equation (85) by chaining two matrixes S_i to two other matrixes.

This classic formulation allowed us to analyze several microwave devices [18–20].

4. Numerical Results and Discussion

4.1. Propagation Modes

Consider TE mode waves in a metallic cylindrical waveguide of radius $R = 13.4$ mm, fully filled with LMF (see Figure 1) with a partial magnetization $4\pi M$. The waveguide has a resonant frequency of 6.57 GHz for the fundamental mode if it is empty. For the case of ferrite magnetized longitudinally with $\frac{M}{M_S} = 0.8$, $4\pi M_S = 750$ G and a permittivity $\varepsilon_{rf} = 11.3$ (ferrite TT1-414 from Trans-Tech), the resonant frequency is $f_{c.11}^{TE} = 3$ GHz for the TE_{11} mode. When the ferrite is demagnetized, the resonant frequency is $f_{c.11}^{TE} = 1.96$ GHz.

Figure 4 represents calculated curves of the propagation constant for the first three TE modes in the frequency range 1–8 GHz and for the cases where the ferrite is magnetized with $\frac{M}{M_S} = 0.8$ or demagnetized. All modes propagate. The propagation constant decreases when the magnetization increases.

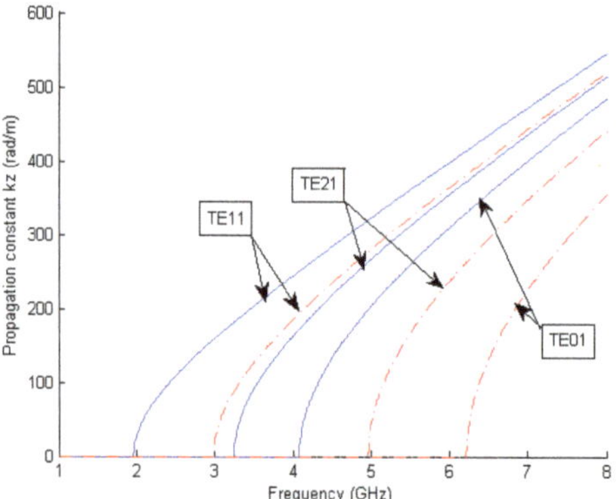

Figure 4. Curves of the propagation constant k_z^{TE} for the first three transverse electric (TE) modes of the cylindrical waveguide completely filled with LMF. ______: case for a demagnetized ferrite. _._._._.: case for $\frac{M}{M_S} = 0.8$.

In Figure 5, the cutoff frequencies of the lowest TE modes are shown as $\frac{M}{M_S}$ increases from 0 to 0.99. It can be seen that the cutoff frequencies of the TE modes monotonically decrease as $\frac{M}{M_S}$ decreases. We can notice the same behavior for the TM modes. The numerical results show that the cutoff frequencies can be controlled by external magnetization.

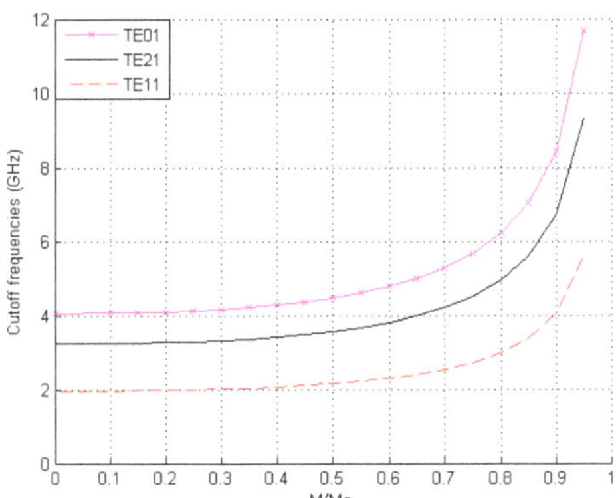

Figure 5. The cutoff frequencies for the first three TE modes vs. $\frac{M}{M_S}$.

4.2. Ferrite Antenna Design

In this section, we describe how the use of MM can be extended to characterize uni-axial discontinuities between cylindrical waveguides filled with the studied media. The discontinuities were considered without losses. This method was based on the modal development of the transverse electromagnetic fields.

We considered two discontinuities (see Figure 3) constituted by juxtaposing three cylindrical waveguides with the same dimensions (R = 13.4 mm). The central waveguide of width d = 5 mm was filled by LMF, as was studied in the previous section. The other guides were empty.

Figure 6 represents the reflection coefficient as a function of the frequency using our approach and HFSS. For the modal method, we used eight modes in the whole circuit. We note that both simulations were in perfect agreement. However, our method was significantly faster than HFSS (7.12 s vs. 9.43 min) because it is a modal method. Thus, by using our approach, it is easy to design antenna according to given specifications.

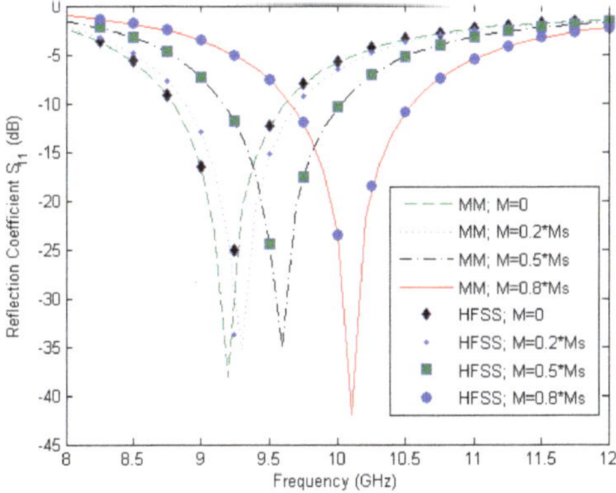

Figure 6. Reflection coefficient of circular waveguide antenna in Figure 3.

The magnetization of the ferrite using an external control can be achieved by enveloping the waveguide with Helmholtz coils and connecting them to a variable voltage generator. Each voltage corresponds to a constant magnetic field applied to the ferrite longitudinally that creates the magnetization of this medium. Our interest in this study was in the tuning of the operating frequency of the antenna.

Figure 7 represents the results of the simulation of the radiation pattern and the gain of antenna with HFSS for various values of magnetized ferrite at the resonance frequency.

For demagnetized ferrite the gain and resonance frequency of this antenna are 8.93 dB and 9.2 GHz, respectively. It is the geometric discontinuity between the empty cylinder and the free air which radiates. However, the physical discontinuities between the guides and between the materials (ferrite–air) have an effect on the radiation of the antenna. The gain of the antenna depends on the electrical and magnetic characteristics of the ferrite medium.

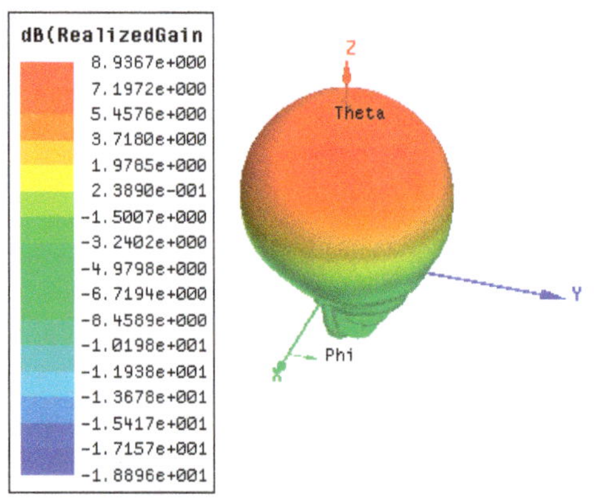

(**a**) Frequency = 9.2 GHz and M = 0xMs

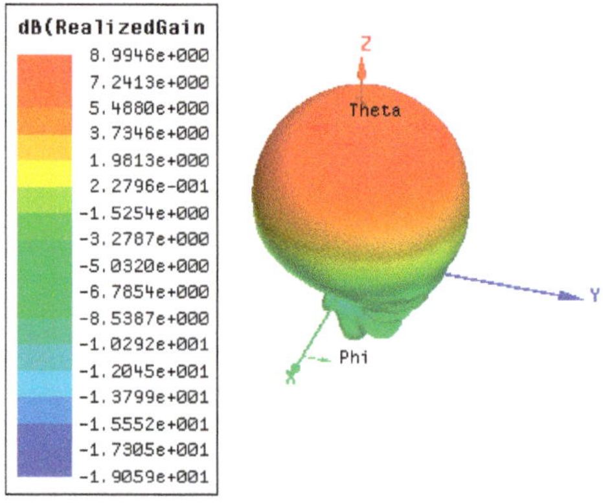

(**b**) Frequency = 9.3 GHz and M = 0.2xMs

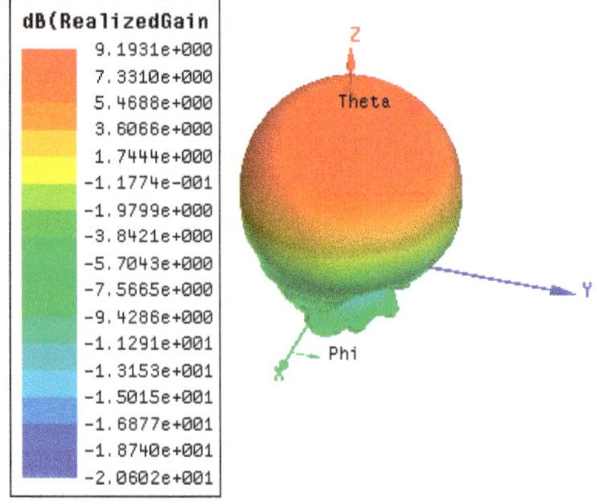

(**c**) Frequency = 9.6 GHz and M = 0.5xMs

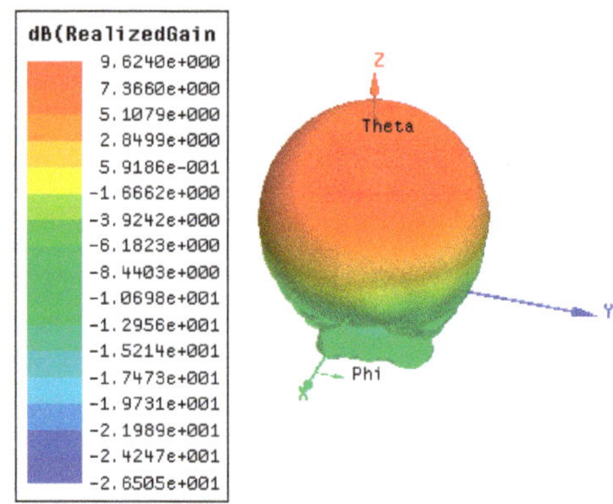

(**d**) Frequency = 10.1 GHz and M = 0.8xMs(**d**)

Figure 7. Radiation pattern and gain of antenna for various values of magnetized ferrite at the resonant frequency.

When the magnetization increases, the gain and the resonant frequency increase. At a magnetization equaling 0.8 Ms, the gain reaches 9.62 dB and the resonant frequency passes to 10.1 GHz.

If magnetization M increases, then the cutoff increases due to the decrease in inter-modal interference. So, the gain increases.

We can externally control the resonant frequency with the tensor permeability, which is a function of the magnetization of ferrite. The resonant frequency increases with an increase of the magnetization. The antenna has become multiband.

5. Conclusions

A rigorous TE and TM modes analysis of cylindrical waveguides completely filled with longitudinally magnetized ferrite was developed in this study. It was demonstrated that the electromagnetic characteristics of the waveguide are closely dependent on magnetization. The curves of the dispersion diagram of the fundamental mode and the first two higher order modes of the ferrite waveguide were obtained. The cutoff frequencies could be controlled by external magnetization. These results for the propagation constant can be used in the design of cylindrical ferrite waveguide antennas. Our results were in good agreement with the theoretical prediction. Moreover, in this paper, we applied the mode matching technique to analyze multiple uni-axial discontinuities in metallic cylindrical waveguides filled with anisotropic materials. The results of the simulations of the cylindrical waveguides antenna filled with partially magnetized ferrite using the mode matching technique and those obtained with HFSS were in perfect agreement. However, our method was noticeably faster. The proposed formulation is a useful tool for microwave engineers.

Funding: This research received no external funding.

Acknowledgments: This work was supported by the Research and Development Division of the society of EITA Consulting in France.

Conflicts of Interest: The author declares no conflict of interest.

References

1. Balanis, C.A. *Circular Waveguides*; Material in Advanced Engineering Electromagnetics; Sect. 9.2; Wiley: New York, NY, USA, 1989; pp. 643–650.
2. Benzina, H.; Sakli, H.; Aguili, T. Complex mode in rectangular waveguide filled with longitudinally magnetized ferrite slab. *Prog. Electromagn. Res. M* **2010**, *11*, 79–87. [CrossRef]
3. Boyenga, D.L.; Mabika, C.N.; Diezaba, A. A New Multimodal Variational Formulation Analysis of Cylindrical Waveguide Uniaxial Discontinuities. *Res. J. Appl. Sci. Eng. Technol.* **2013**, *6*, 787–792. [CrossRef]
4. Dong, J.F.; Li, J. Characteristics of guided modes in uniaxial chiral circular waveguides. *Prog. Electromagn. Res. Pier.* **2012**, *124*, 331–345. [CrossRef]
5. Green, J.J.; Sand, F. Microwave characterization of partially magnetized ferrites. *IEEE Trans. Microw. Theory Tech.* **1974**, *22*, 641–645. [CrossRef]
6. Mahmoud, S.F. Guided modes on open chirowaveguides. *IEEE Trans. Microw. Theory Tech.* **1995**, *43*, 205–209. [CrossRef]
7. Meng, F.Y.; Wu, Q.; Li, L.W. Controllable Metamaterial-Loaded Waveguides Supporting Backward and Forward Waves Transmission characteristics of wave modes in a rectangular waveguide filled with anisotropic metamaterial. *IEEE Trans. Antennas Propag.* **2011**, *59*, 3400–3411. [CrossRef]
8. Polder, D. On the theory of ferromagnetic resonance. *Philos. Mag.* **1949**, *40*, 99–115. [CrossRef]
9. Schloemann, E. Microwave behavior of partially magnetized ferrites. *J. Appl. Phys.* **1970**, *1*, 1204.
10. Thabet, R.; Riabi, M.L.; Belmeguenai, M. Rigorous design and efficient optimization of quarter-wave transformers in metallic circular waveguides using the mode-matching method and the genetic algorithm. *Prog. Electromagn. Res. PIER* **2007**, *68*, 15–33.
11. Sakli, H.; Benzina, H.; Aguili, T.; Tao, J.W. Propagation constant of a rectangular waveguides completely full of ferrite magnetized longitudinally. *IJIM–Springer Int. J. Infrared Millim. Waves* **2009**, *30*, 877–883. [CrossRef]
12. Zeller, G.P.; McFarland, K.S.; Adams, T.; Alton, A.; Avvakumov, S.; de Barbaro, L.; de Barbaro, P.; Bernstein, B.H.; Bodek, A.; Bolton, T.; et al. Erratum: Precise determination of electroweak parameters in neutrino-nucleon scatterin. *Phys. Rev. Lett.* **2002**, *89*, 183901.
13. Shadrivov, I.V.; Sukhorukov, A.A.; Kivshar, Y.S. Guided modes in negative-refractive-index waveguides. *Phys. Rev. E* **2003**, *67*, 057602. [CrossRef] [PubMed]
14. Marvasti, M.; Rejaei, B. Formation of hotspots in partially filled ferrite-loaded rectangular waveguides. *J. Appl. Phys.* **2017**, *122*, 233901. [CrossRef]
15. Mitu, S.S.I.; Al–Garni, A.M.; Ragheb, H.A. Ferrite–Loaded Circular Waveguide Antenne for 3D Scanning. U.S. Patent No. US 9 979 085 B2, 22 May 2018.
16. Couffignal, P. Contribution à L'étude Des Filtres en Guides Métalliques. Ph.D. Thesis, INP Toulouse, Toulouse, France, 1992.
17. Tao, J.W. Travaux Scientifiques, Habilitation à Diriger Des Recherches. Ph.D. Thesis, Université de Savoie, Chambéry, France, 1999.
18. Tao, J.W.; Baudrand, B. Multimodal variational analysis of uniaxial waveguide discontinuities. *IEEE Trans. Microw. Theory Tech.* **1991**, *39*, 506–516. [CrossRef]
19. Lilonga, D.; Tao, J.W.; Vuong, T.H. Uniaxial discontinuities analysis by a new multimodal variational method: Application to filter design. *Int. J. Rf Microw. Comput. -Aided Eng.* **2006**, *17*, 77–83. [CrossRef]
20. Yahia, M.; Tao, J.W.; Sakli, H. Complex Rectangular Filter Design Using Hybrid Finite Element Method and Modified Multimodal Variational Formulation. *Pier C Prog. Electromagn. Res. C* **2013**, *44*, 55–66. [CrossRef]

 *electronics*

Article

Multiple Scattering by Two PEC Spheres Using Translation Addition Theorem

Sidra Batool [1,*], Mehwish Nisar [1], Lorenzo Dinia [1], Fabio Mangini [2] and Fabrizio Frezza [1]

[1] Department of Information Engineering, Electronics, and Telecommunications, Sapienza University of Rome, Via Eudossiana 18, 00184 Rome, Italy; mehwish.nisar@uniroma1.it (M.N.); lorenzo.dinia@uniroma1.it (L.D.); fabrizio.frezza@uniroma1.it (F.F.)

[2] Department of Information Engineering, University of Brescia, Via Branze 59, 25123 Brescia, Italy; fabio.mangini@unibs.it

* Correspondence: sidra.batool@uniroma1.it

Abstract: An analysis of multiple scattering by two Perfect Electric Conducting (PEC) spheres using translation Addition Theorem (AT) for spherical vector wave functions is presented. Specifically, the Cruzan formalism is used to represent the AT for spherical harmonics, which introduces the translation coefficients for transformation of spherical harmonics from one coordinate to another. The adoption of these coefficients with the use of two PEC spheres in a near zone region makes the calculation of multiple scattering electric fields very efficient. As an illustration, the mathematical formation using advanced computational approaches was inspected. Then, the generic truncation criteria in the scattered electric field by two PEC spheres was deeply investigated using translation AT. However, the numerical validation was obtained using Comsol simulation software. This approach will allow to evaluate the scattering from macro-structures composed of spherical particles, i.e., biological molecules, clouds of airborne particles, etc. An original and fully general solution to the problem using vector quantities is introduced, and the convergence of the solution in several numerical examples is also demonstrated. This approach takes into account the effect of multiple scattering by two PEC spheres for spherical vector function.

Keywords: scattering; translation addition theorem; Mie theory

Citation: Batool, S.; Nisar, M.; Dinia, L.; Mangini, F.; Frezza, F. Multiple Scattering by Two PEC Spheres Using Translation Addition Theorem. *Electronics* **2022**, *11*, 126. https://doi.org/10.3390/electronics11010126

Academic Editor: Gerardo Di Martino

Received: 18 November 2021
Accepted: 25 December 2021
Published: 31 December 2021

Publisher's Note: MDPI stays neutral with regard to jurisdictional claims in published maps and institutional affiliations.

1. Introduction

The problem of multiple scattering by closely spaced objects has a wide range of engineering applications, including electromagnetic (EM) wave transmission by rain [1], scattering by complex bodies [2–5], scanning of buried objects [6,7], biological cell detection [8,9], radar and remote sensing applications in biomedical diagnostics, etc. [10,11]. The resolution of multiple scattering problems by PEC spheres allows an analytic treatment and a better physical interpretation into the scattering mechanism for novel applications [12,13].

The problem of scattering from two identical spheres with small radii was formulated by Liang and Lo [14]. The translation AT for vector spherical wave functions as multipole expansion was used to express the derived solution of EM fields distributed by spheres. Later on, Olaofe [15] described the multiple scattering by an unequal and parallel circular cylinders. An overview of previous research projects on multiple scattering problems showed that a theoretical investigation of the effects of inter-particle coupling on morphology-dependent resonances of spheres was examined by Fuller [16]. Lo and Bruning determined the new recursion relationship for the calculations of the multiple scattering of EM waves by two arbitrary spheres, reducing the difficulty of the computational quantitative analysis in scattering problems [17]. Further, Wang and Chew derived the recursive approach (T-matrix algorithm), which is used for the formation of multiple scattering fields by several spheres. This method is suitable for the calculation of the vector AT and valid for Monte Carlo simulation for many diagnostic applications [18].

In 1954, Friedman and Russek proposed the calculation of AT for spherical scalar wave functions [19]. Cruzen and Stein also developed the AT for spherical harmonics, fitting it for the appropriate solution of multiple scattering by spheres [20,21]. More recently, a further improvement was introduced by Mackowski with the superposition solution method for multiple spheres scattering problems [22]. Xu illustrated the multiple scattering by aggregate spheres, and interpreted the detail of the fast evaluation of Gaunt coefficients [23]. The vector translation AT for spherical harmonics was first formulated by Xu for an analytical solution of multiple scattering problems. Then, the computational difference between the three vital types of analytical expressions was determined for the vector translation coefficients: Stein's [20], Cruzan's [21], and Xu's [24]. Moreover, Xu presented the necessary recursive approaches to exactly calculating the Gaunt coefficients using Wigner 3-j [25].

Recently, Batool et al. specified a brief outline of the multiple scattering by a PEC sphere using translation AT for spherical harmonics [26]. Additionally, the effect of truncation error on its convergence behavior in AT for spherical harmonics was deduced [27]. As is known, some researchers reported the truncation error and its convergence in Mie theory [28]. It was observed that the most popular Wiscombe's criterion was not sufficient to overcome the truncation error of translation AT for spherical vector functions [29].

In the current manuscript, the multiple scattering analysis using two PEC spheres was studied. The Cruzen formula for the translation addition coefficients based on Wigner 3-j was selected. Then, the scattering electric field using AT for spherical harmonics was derived and the mathematical formation using numerical simulation approach was computed. During our numerical investigation, the truncation error and its convergence were observed to vary with respect to the frequency. The study offers an examination of the impact of finite terms on the truncation error of AT related to spherical vector wave function. The truncation effect forces the numerical results of spherical harmonics to vary, as common for the most truncation criteria used. Currently, the most used truncation criterion in the literature is the one proposed by Wiscombe more than 40 years ago. This criterion allows to accurately choose the truncation of the series in order to express an electric field in the presence of a single scatterer. On the other hand, it does not take into consideration more complex scenarios, as in the case of two or more scatterers. In this scenario, the field must be expressed in spherical vector wave functions translated on the center of a reference system of each other sphere. This translation is reached by exploiting the AT, which in turn is obtained by a superposition of vector spherical functions that must obey an appropriate truncation criterion. As a consequence of the numerical simulation results, the translation of both spheres along z-axis and the observed scattering pattern by varying frequency and radii of the spheres were presented. Comsol (Multiphysics 5.4) simulation software for the validation of the numerical results was used and the best comparison between Matlab and Comsol results was also retrieved.

2. Formation of the Problem

Two PEC spheres with radii a and c with respect to the center at the origins O and O' of two different coordinates systems are taken under consideration. The sphere with O as origin is characterized by a spherical coordinates system (r, θ, ϕ), while (r', θ', ϕ') characterizes the sphere with origin O'. The distance between the centers of spheres along z-axis is δ as shown in Figure 1. Let us study a horizontal elliptically polarized plane wave that propagates along a certain direction of the three dimensions cartesian coordinates system. The plane wave can be written as [30,31]:

$$\mathbf{E}_i(\mathbf{r}) = \mathbf{e}_{pol} \exp(i\mathbf{k}_i\mathbf{r}) = \left(E_\theta \boldsymbol{\theta} + E_\phi \boldsymbol{\phi}\right) \exp(i\mathbf{k}_i\mathbf{r}) \tag{1}$$

with

$$\mathbf{k}_i = k_1(\sin\theta_i\cos\phi_i\mathbf{x_0} + \sin\theta_i\sin\phi_i\mathbf{y_0} + \cos\theta_i\mathbf{z_0}) \tag{2}$$

$$\boldsymbol{\theta}_{0i} = (\cos\theta_i\cos\phi_i\mathbf{x_0} + \cos\theta_i\sin\phi_i\mathbf{y_0} - \sin\theta_i\mathbf{z_0}) \tag{3}$$

$$\boldsymbol{\phi}_{0i} = -\sin\phi_i\mathbf{x_0} + \cos\phi_i\mathbf{y_0} \tag{4}$$

Then, the incident electric field in terms of spherical harmonics can be written as [30,31]:

$$\mathbf{E}_i(\mathbf{r}) = \sum_{\nu=1}^{+\infty}\sum_{\mu=-\nu}^{\nu}\left[a_{\mu\nu}\mathbf{M}^{(1)}_{\mu\nu}(\mathbf{r}) + b_{\mu\nu}\mathbf{N}^{(1)}_{\mu\nu}(\mathbf{r})\right] \tag{5}$$

Vector spherical harmonics $\mathbf{M}^{(1)}_{\mu\nu}$ and $\mathbf{N}^{(1)}_{\mu\nu}$ is [30,31]:

$$\mathbf{M}^{(1)}_{\mu\nu} = \exp(i\mu\phi)j_\nu(kr)\left[i\mu\frac{P^\mu_\nu(\cos\theta)}{\sin\theta}\boldsymbol{\theta}_0 - \frac{\partial P^\mu_\nu(\cos\theta)}{\partial\theta}\boldsymbol{\phi}_0\right] \tag{6}$$

$$\mathbf{N}^{(1)}_{\mu\nu} = \exp(i\mu\phi)\frac{j_\nu(kr)}{kr}\nu(\nu+1)P^\mu_\nu(\cos\theta)\mathbf{r}_0+ \tag{7}$$

$$\left[\frac{\partial P^\mu_\nu(\cos\theta)}{\partial\theta}\boldsymbol{\theta}_0 + i\frac{\mu P^\mu_\nu(\cos\theta)}{\sin\theta}\boldsymbol{\phi}_0\right]\exp(i\mu\phi)\frac{1}{kr}\frac{\partial}{\partial r}[rj_\nu(kr)]$$

considering

$$\pi^\mu_\nu(\cos\theta) = \mu\frac{P^\mu_\nu(\cos\theta)}{\sin\theta} \tag{8}$$

$$\tau^\mu_\nu(\cos\theta) = \frac{\partial P^\mu_\nu(\cos\theta)}{\partial\theta} \tag{9}$$

The vector spherical harmonics expressions can be simplified as follows:

$$\mathbf{M}^{(1)}_{\mu\nu} = \exp(i\mu\phi)j_\nu(kr)\left[i\pi^\mu_\nu(\cos\theta)\boldsymbol{\theta}_0 - \tau^\mu_\nu(\cos\phi)\boldsymbol{\phi}_0\right] \tag{10}$$

$$\mathbf{N}^{(1)}_{\mu\nu} = \exp(i\mu\phi)\frac{j_\nu(kr)}{kr}\nu(\nu+1)P^\mu_\nu(\cos\theta)\mathbf{r}_0+ \tag{11}$$

$$\left[\tau^\mu_\nu(\cos\theta)\boldsymbol{\theta}_0 + i\pi^\mu_\nu(\cos\phi)\boldsymbol{\phi}_0\right]\exp(i\mu\phi)\frac{1}{kr}\frac{\partial}{\partial r}[rj_\nu(kr)]$$

Now, let us consider modified spherical vector functions characterized by an angles θ and ϕ. The following expressions are obtained replacing the spherical vector function with the tesseral function:

$$\mathbf{m}_{\mu\nu} = \exp(i\mu\phi)\left[i\pi^\mu_\nu(\cos\theta)\boldsymbol{\theta}_0 - \tau^\mu_\nu(\cos\theta)\boldsymbol{\phi}_0\right] \tag{12}$$

$$\mathbf{n}_{\mu\nu} = \exp(i\mu\phi)\left[\tau^\mu_\nu(\cos\theta)\boldsymbol{\theta}_0 + i\pi^\mu_\nu(\cos\theta)\boldsymbol{\phi}_0\right] \tag{13}$$

$$\mathbf{p}_{\mu\nu} = \exp(i\mu\phi)\nu(\nu+1)P^\mu_\nu(\cos\theta)\mathbf{r}_0 \tag{14}$$

By above-simplified Equations (10) and (11), we achieve:

$$\mathbf{M}^{(1)}_{\mu\nu} = j_\nu(kr)\mathbf{m}_{\mu\nu} \tag{15}$$

$$\mathbf{N}^{(1)}_{\mu\nu} = \frac{j_\nu(kr)}{kr}\mathbf{p}_{\mu\nu} + \frac{1}{kr}\frac{\partial}{\partial r}[rj_\nu(kr)]\mathbf{n}_{\mu\nu} \tag{16}$$

The features of orthogonality properties are used for vector spherical harmonics, which implies that the coefficients can be expressed in a modified form:

$$a_{\mu\nu} = i^{\nu} \frac{(2\nu+1)(\nu-\mu)!}{\nu(\nu+1)(\nu+\mu)!} \mathbf{e}_{pol} \cdot \mathbf{m}_{\mu\nu}^{*}(\theta_t, \phi_t) \tag{17}$$

$$b_{\mu\nu} = i^{\nu-1} \frac{(2\nu+1)(\nu-\mu)!}{\nu(\nu+1)(\nu+\mu)!} \mathbf{e}_{pol} \cdot \mathbf{n}_{\mu\nu}^{*}(\theta_t, \phi_t) \tag{18}$$

The elliptically polarized incident field for a sphere may once be written as:

$$\mathbf{E}_i(r,\theta,\phi) = \sum_{\nu=1}^{+\infty} \sum_{\mu=-\nu}^{\nu} \left[a_{\mu\nu}\mathbf{M}_{\mu\nu}^{(1)}(r,\theta,\phi) + b_{\mu\nu}\mathbf{N}_{\mu\nu}^{(1)}(r,\theta,\phi) \right] \tag{19}$$

with

$$a_{\mu\nu} = i^{\nu} \frac{(2\nu+1)(\nu-\mu)!}{\nu(\nu+1)(\nu+\mu)} \exp\left(-i\mu\phi_i\right) \left[-iE_{\theta_i}\pi_{\nu}^{\mu}(\cos\theta_i) - E_{\phi_i}\tau_{\nu}^{\mu}(\cos\theta_i) \right] \tag{20}$$

$$b_{\mu\nu} = i^{\nu-1} \frac{(2\nu+1)(\nu-\mu)!}{\nu(\nu+1)(\nu+\mu)} \exp\left(-i\mu\phi_i\right) \left[-iE_{\theta_i}\tau_{\nu}^{\mu}(\cos\theta_i) - E_{\phi_i}\pi_{\nu}^{\mu}(\cos\theta_i) \right] \tag{21}$$

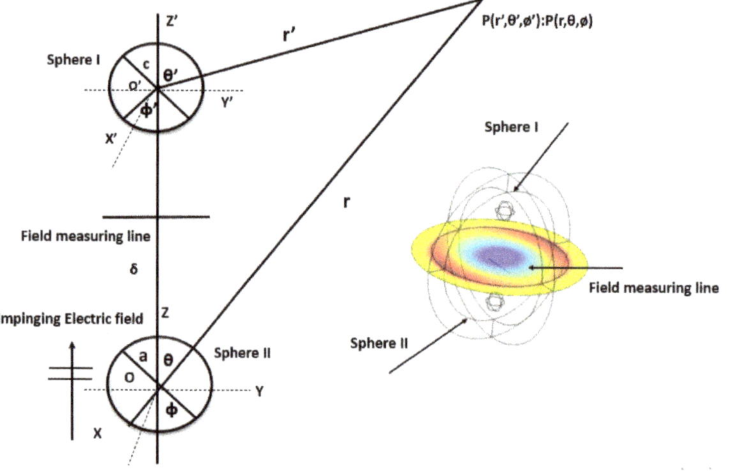

Figure 1. Geometry of the two PEC spheres exercising translation of the spheres along-z-axis using translation AT.

2.1. Expansion of Incident Plane Wave

The incident electric field is expanded into multipole fields series around the origins O and O' of the spherical coordinates systems (r,θ,ϕ) and (r',θ',ϕ'). Thus, the incident plane wave using multipole coefficients expansion may be written as follows, since the radial vector $r = r' + \delta = \exp(-ik_1 r')\exp(-ik_1\delta\cos\alpha)$ [14].

$$\mathbf{E}_i(r',\theta',\phi') = \sum_{\nu=1}^{+\infty} \sum_{\mu=-\nu}^{\nu} \left[a_{\mu\nu}^{*}\mathbf{M}_{\mu\nu}^{(1)}(r',\theta',\phi') + b_{\mu\nu}^{*}\mathbf{N}_{\mu\nu}^{(1)}(r',\theta',\phi') \right] \tag{22}$$

where

$$a_{\mu\nu}^{*} = \exp(-ik_1\delta\cos\alpha)a_{\mu\nu} \tag{23}$$

$$b_{\mu\nu}^{*} = \exp(-ik_1\delta\cos\alpha)b_{\mu\nu} \tag{24}$$

2.2. Scattered Field by Two Conducting Sphere

Scattered field by two PEC spheres may be represented with $\mathbf{E}_s^I$ and $\mathbf{E}_s^{II}$, respectively. In this problem, the total electric field is described by the incident and scattered electric fields, and it is expressed as follows:

$$\mathbf{E}^{total} = \mathbf{E}_i + \mathbf{E}_s^I + \mathbf{E}_s^{II} \tag{25}$$

where

$$\mathbf{E}_s^I(r,\theta,\phi) = \sum_{v=1}^{+\infty} \sum_{\mu=-v}^{v} \left[e_{\mu v} \mathbf{M}_{\mu v}^{(3)}(r,\theta,\phi) + f_{\mu v} \mathbf{N}_{\mu v}^{(3)}(r,\theta,\phi) \right] \tag{26}$$

$$\mathbf{E}_s^{II}(r',\theta',\phi') = \sum_{v=1}^{+\infty} \sum_{\mu=-v}^{v} \left[g_{\mu v} \mathbf{M}_{\mu v}^{(3)}(r',\theta',\phi') + h_{\mu v} \mathbf{N}_{\mu v}^{(3)}(r',\theta',\phi') \right] \tag{27}$$

2.3. AT for Translation of the Vector Spherical Harmonics

The translation AT for spherical vector wave equations depends on the relative schism in a spherical coordinates system using different origin O and O'. The following expressions are used to express the translation of l_{th} coordinates system (r,θ,ϕ) to j_{th} (r',θ',ϕ') coordinates system.

$$\mathbf{M}_{\mu v}^{(3)}(l) = \sum_{n=1}^{\infty} \sum_{m=-n}^{n} A_{mn}^{\mu v}(l,j)\mathbf{M}_{mn}^{(1)}(j) + B_{mn}^{\mu v}(l,j)\mathbf{N}_{mn}^{(1)}(j) \tag{28}$$

$$\mathbf{N}_{\mu v}^{(3)}(l) = \sum_{n=1}^{\infty} \sum_{m=-n}^{n} B_{mn}^{\mu v}(l,j)\mathbf{M}_{mn}^{(1)}(j) + A_{mn}^{\mu v}(l,j)\mathbf{N}_{mn}^{(1)}(j) \tag{29}$$

Then, translation of a jth coordinates system (r',θ',ϕ') to lth coordinates system (r,θ,ϕ) at translation distance δ may be formulated as:

$$\mathbf{M}_{\mu v}^{(3)}(j) = \sum_{n=1}^{\infty} \sum_{m=-n}^{n} C_{mn}^{\mu v}(j,l)\mathbf{M}_{mn}^{(1)}(l) + D_{mn}^{\mu v}(j,l)\mathbf{N}_{mn}^{(1)}(l) \tag{30}$$

$$\mathbf{N}_{\mu v}^{(3)}(j) = \sum_{n=1}^{\infty} \sum_{m=-n}^{n} D_{mn}^{\mu v}(j,l)\mathbf{M}_{mn}^{(1)}(l) + C_{mn}^{\mu v}(j,l)\mathbf{N}_{mn}^{(1)}(l) \tag{31}$$

2.4. Formation of Vector Translation Coefficients

EM field in terms of translation is illustrated by an infinite sum with respect to a coordinate system with a different reference. Therefore, Cruzan's mathematical derivations for $A_{mn\mu v}^{l,j}$ and $B_{mn\mu v}^{l,j}$ coefficients can be written in a simplified form [21]:

$$\begin{aligned} A_{mn\mu v}^{l,j} = (-1)^{-m} & \frac{(2v+1)(n+m)!(v-\mu)!}{2n(n+1)(n-m)!(v+\mu)!} \exp[i(\mu-m)\phi_{lj}] \\ & \times \sum_{q=0}^{q_{max}} i^p [n(n+1) + v(v+1) - p(p+1)] a_q \\ & \times h_p^{(1)}(kd) P_p^{\mu-m}(\cos\theta) \end{aligned} \tag{32}$$

The following expressions are reached:

$$A_{mn\mu v}^{l,j} = (-1)^{-m} \zeta(m,n,\mu,v) \sum_{q=0}^{q_{max}} h_p^{(1)}(kr) \tag{33}$$

$$C_{mn\mu v}^{l,j} = (-1)^{-m} \zeta^*(m,n,\mu,v) \sum_{q=0}^{q_{max}} h_p^{(1)}(kr) \tag{34}$$

where

$$\zeta(m,n,\mu,\nu) = \frac{(2\nu+1)(n+m)!(\nu-\mu)!}{2n(n+1)(n-m)!(\nu+\mu)!} \exp[i(\mu-m)\phi_{lj}]$$

$$\times \sum_{q=0}^{q_{max}} i^p [n(n+1) + \nu(\nu+1) - p(p+1)] a_q \quad P_p^{\mu-m}(\cos\theta_{lj})$$

and

$$B_{mn\mu\nu}^{l,j} = (-1)^{-m+1} \frac{(2\nu+1)(n+m)!(\nu-\mu)!}{2n(n+1)(n-m)!(\nu+\mu)!} \exp[i(\mu-m)\phi_{lj}]$$

$$\times \sum_{q=1}^{Q_{max}} i^{p+1} \{[(p+1)^2 - (n-\nu)^2][(n+\nu+1)^2 - (p+1)^2]\}^{\frac{1}{2}} \tag{35}$$

$$\times b(-m,n,\mu,\nu,p+1,p) h_{p+1}^{(1)}(kr) P_{p+1}^{\mu-m}(\cos\theta)$$

Similarly

$$B_{mn\mu\nu}^{l,j} = (-1)^{-m+1} \zeta(m,n,\mu,\nu) \sum_{q=0}^{q_{max}} h_{p+1}^{(1)}(kr) \tag{36}$$

$$D_{mn\mu\nu}^{l,j} = (-1)^{-m+1} \zeta^*(m,n,\mu,\nu) \sum_{q=0}^{q_{max}} h_{p+1}^{(1)}(kr) \tag{37}$$

where

$$\xi(m,n,\mu,\nu) = \frac{(2\nu+1)(n+m)!(\nu-\mu)!}{2n(n+1)(n-m)!(\nu+\mu)!} \exp[i(\mu-m)\phi_{lj}]$$

$$\times \sum_{q=1}^{Q_{max}} i^{p+1} \{[(p+1)^2 - (n-\nu)^2][(n+\nu+1)^2 - (p+1)^2]\}^{\frac{1}{2}} \tag{38}$$

$$\times b(-m,n,\mu,\nu,p+1,p) P_{p+1}^{\mu-m}(\cos\theta_{lj})$$

Similarly, $C_{mn\mu\nu}^{j,l}$ and $D_{mn\mu\nu}^{j,l}$ coefficients are obtained by taking the complex conjugate of $\xi(m,n,\mu,\nu)$ and $\zeta(m,n,\mu,\nu)$ from $A_{mn\mu\nu}^{j,l}$ and $B_{mn\mu\nu}^{j,l}$. Here, k is the propagation constant, $a_q = a(-m,n,\mu,\nu,p)$, $q = 1, 2, \ldots, q_{max}$, $p = n + \nu - 2q$ and

$$q_{max} = min\left(n, \nu, \frac{n+\nu-|m+\mu|}{2}\right) \tag{39}$$

Gaunt coefficients have been described in the literature [24], the total electric field for the *l*th coordinates system (r, θ, ϕ) may be explicitly written as:

$$\mathbf{E}_{total}(l) = \sum_{\nu=1}^{+\infty} \sum_{\mu=-\nu}^{\nu} a_{\mu\nu} \mathbf{M}_{\mu\nu}^{(1)}(l) + b_{\mu\nu} \mathbf{N}_{\mu\nu}^{(1)}(l) + e_{\mu\nu} \mathbf{M}_{\mu\nu}^{(3)}(l) + f_{\mu\nu} \mathbf{N}_{\mu\nu}^{(3)}(l)$$

$$+ \sum_{\nu=1}^{+\infty} \sum_{\mu=-\nu}^{\nu} \sum_{n=1}^{\infty} \sum_{m=-n}^{n} g_{\mu\nu} \left\{ A_{mn}^{\mu\nu}(l,j) \mathbf{M}_{mn}^{(1)}(l) + B_{mn}^{\mu\nu}(l,j) \mathbf{N}_{mn}^{(1)}(l) \right\} + \tag{40}$$

$$h_{\mu\nu} \left\{ B_{mn}^{\mu\nu}(l,j) \mathbf{M}_{mn}^{(1)}(l) + A_{mn}^{\mu\nu}(l,j) \mathbf{N}_{mn}^{(1)}(l) \right\}$$

Similarly, the total electric field for the j_{th} coordinates system (r', θ', ϕ') may be written as:

$$\mathbf{E}_{total}(j) = \sum_{v=1}^{+\infty} \sum_{\mu=-v}^{v} a_{\mu v}^* \mathbf{M}_{\mu v}^{(1)}(j) + b_{\mu v}^* \mathbf{N}_{\mu v}^{(1)}(j) + g_{\mu v} \mathbf{M}_{\mu v}^{(3)}(j) + h_{\mu v} \mathbf{N}_{\mu v}^{(3)}(j)$$
$$+ \sum_{v=1}^{+\infty} \sum_{\mu=-v}^{v} \sum_{n=1}^{\infty} \sum_{m=-n}^{n} e_{\mu v} \left\{ C_{mn}^{\mu v}(j,l) \mathbf{M}_{mn}^{(1)}(j) + D_{mn}^{\mu v}(j,l) \mathbf{N}_{mn}^{(1)}(j) \right\} + \tag{41}$$
$$f_{\mu v} \left\{ D_{mn}^{\mu v}(j,l) \mathbf{M}_{mn}^{(1)}(j) + C_{mn}^{\mu v}(j,l) \mathbf{N}_{mn}^{(1)}(j) \right\}$$

Applying the boundary condition, the tangential component of the electric field must continue on the surface of spheres. Using the orthogonality properties, the simultaneous linear equations can be expressed as:

$$e_{\mu v} = X_n(a) \left\{ a_{\mu v} + \left(g_{\mu v} \sum_{n=1}^{\infty} \sum_{m=-n}^{n} A_{mn}^{\mu v} + h_{\mu v} \sum_{n=1}^{\infty} \sum_{m=-n}^{n} B_{mn}^{\mu v} \right) \right\} \tag{42}$$

$$f_{\mu v} = Y_n(a) \left\{ b_{\mu v} + \left(g_{\mu v} \sum_{n=1}^{\infty} \sum_{m=-n}^{n} B_{mn}^{\mu v} + h_{\mu v} \sum_{n=1}^{\infty} \sum_{m=-n}^{n} A_{mn}^{\mu v} \right) \right\} \tag{43}$$

$$g_{\mu v} = X_n(c) \left\{ a_{\mu v}^* + \left(e_{\mu v} \sum_{n=1}^{\infty} \sum_{m=-n}^{n} C_{mn}^{\mu v} + f_{\mu v} \sum_{n=1}^{\infty} \sum_{m=-n}^{n} D_{mn}^{\mu v} \right) \right\} \tag{44}$$

$$h_{\mu v} = Y_n(c) \left\{ b_{\mu v}^* + \left(e_{\mu v} \sum_{n=1}^{\infty} \sum_{m=-n}^{n} D_{mn}^{\mu v} + f_{\mu v} \sum_{n=1}^{\infty} \sum_{m=-n}^{n} C_{mn}^{\mu v} \right) \right\} \tag{45}$$

where

$$X(r_0) = -\frac{j_v(kr)}{h_v^{(1)}(kr)} \Big|_{r=r_0} \tag{46}$$

$$Y(r_0) = -\frac{j_v'(kr)}{h_v'^{(1)}(kr)} \Big|_{r=r_0} \tag{47}$$

After solving the linear equations, all coefficients $e_{\mu v}, f_{\mu v}, g_{\mu v}, h_{\mu v}$ can be obtained, thanks to which it is easily possible to determine scattering, extinction, and absorption cross-section [32]. As a result of the above derivations and the use of Equations (28) and (29), the scattered field can be achieved:

$$\mathbf{E}_s(l) = \sum_{v=1}^{\infty} \sum_{\mu=-v}^{v} \left[e_{\mu v} \sum_{n=1}^{\infty} \sum_{m=-n}^{n} \left\{ A_{mn}^{\mu v}(l,j) \mathbf{M}_{mn}^{(3)}(j) + B_{mn}^{\mu v}(l,j) \mathbf{N}_{mn}^{(3)}(j) \right\} \right.$$
$$\left. + f_{\mu v} \sum_{n=1}^{\infty} \sum_{m=-n}^{n} \left\{ B_{mn}^{\mu v}(l,j) \mathbf{M}_{mn}^{(3)}(j) + A_{mn}^{\mu v}(l,j) \mathbf{N}_{mn}^{(3)}(j) \right\} \right] \tag{48}$$

3. Numerical Results

This paper introduces the numerical results of multiple scattering by two PEC spheres. The validity of the present mathematical formalism has been investigated by multi-paradigm programming language (Matlab) and simulation software (Comsol Multiphysics 5.4). The proposed method efficiency has been analyzed by comparison of Matlab and Comsol numerical results. This article is dedicated to a scattering theory using translation AT for spherical harmonic functions. Specifically, the calculation of the translation coefficients for two PEC spheres and the numerical outcomes of the scattered electric field between two PEC spheres in the near zone region are discussed. Therefore, Equation (48) was used for numerical simulation using a Matlab code.

3.1. Validation Test for Two PEC Spheres at Zero Translation Distance

In the first numerical test, the translation distance between two PEC spheres is zero. This means that the problem of two PEC spheres is reduced into a problem with a single PEC sphere. Then, the single PEC sphere present in a free space is investigated to demonstrate the validity of numerical simulations. For a single PEC sphere, the scattering coefficients $e_{\mu\nu}$ and $f_{\mu\nu}$ can be obtained by using the solution of two linear equations. As a result of the above calculations $g_{\mu\nu} = 0$ and $h_{\mu\nu} = 0$, and the use of Equations (42) and (43), the scattered coefficients can be achieved:

$$e_{\mu\nu} = X_n(a)a_{\mu\nu} \tag{49}$$

$$f_{\mu\nu} = Y_n(a)b_{\mu\nu} \tag{50}$$

It occurs when the origin of the l_{th} and j_{th} coordinates system are overlapped (i.e., no translation is involved). $m = \mu, n = \nu, A_{mn}^{\mu\nu} \equiv 1, B_{mn}^{\mu\nu} \equiv 0$, and $m \neq \mu, n \neq \nu, A_{mn}^{\mu\nu} \equiv 0$, and $B_{mn}^{\mu\nu} \equiv 0$. From the previous conditions, the following expressions are obtained $\mathbf{M}_{\mu\nu}^{(3)}(l) \equiv \mathbf{M}_{\mu\nu}^{(3)}(j), \mathbf{N}_{\mu\nu}^{(3)}(l) \equiv \mathbf{N}_{\mu\nu}^{(3)}(j)$. Consequently, the scattered field by using the above Equations (49) and (50) may be written as:

$$\mathbf{E}_s(l) = \sum_{\nu=1}^{\infty} \sum_{\mu=-\nu}^{\nu} \left[e_{\mu\nu}\mathbf{M}_{\mu\nu}^{(3)}(l) + f_{\mu\nu}\mathbf{N}_{\mu\nu}^{(3)}(l) \right] \tag{51}$$

For this purpose, the following parameters are selected: radius of the sphere $a = 0.1$ m, frequency $f = 0.3$ GHz, angles of incident $\theta_i = 0$ rad, and $\phi_i = 1$ rad. Considering the position of cartesian coordinates system with center $x_r = y_r = z_r = 0$ and center of the sphere $x_q = 0, y_q = 0, z_q = 0$, the field was measured along the line segment with double of length for the radius of the sphere lying parallel along x-axis, which is placed at hight $z = 0.3$ m, $y = 0$. The numerical results (Matlab code) are comparable with the simulation results (COMSOL Multiphysics 5.4) as shown in Figure 2.

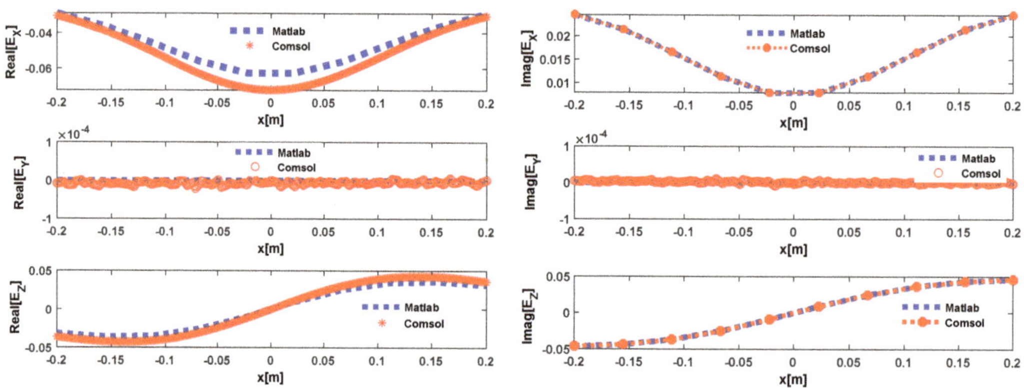

Figure 2. Real part (**left**) and imaginary part (**right**) of the scattered electric field by a PEC sphere present in free space. Matlab results (blue color dot lines) are super-imposed on Comsol results (red marker color lines). Here, selected parameters are frequency $f = 0.3$ GHz, radius $a = c = 0.1$ m.

Further, the scattering behavior with translation of both PEC spheres at different translation distance was investigated. The plots for the scattering electric field are analyzed at three different range of frequencies: 3 MHz, 0.3 GHz, and 3 GHz, respectively. The derived mathematical formalism reveals the response of the truncation error for translation AT of spherical harmonics. It is observed that for lower frequency, the numerical results of the Matlab code presents the optimized convergence at lower truncation sum N, whereas the higher the frequency, the higher the value of truncation sum N for the numerical results of the Matlab code.

3.2. Case 1: f = 3 MHz

In this section, the numerical results of scattered electric fields have been investigated at a frequency $f = 3$ MHz. Let suppose that the PEC spheres I, II were translated along z-axis at $z = 0.3$ m and $z = -0.3$ m, respectively. The position of the field measuring line can be defined as: the line segment to double the radius of sphere that lies parallel along x-axis; that is placed at height $z = 0.1$ m, $y = 0$. For the analysis of the scattering field pattern, the following parameters have been chosen: radius of spheres I, II $a = c = 0.1$ m, angles of incident $\theta_i = 0$ rad, and $\phi_i = 1$ rad. The best comparison between Matlab code and Comsol simulation results are illustrated in Figure 3. The numerical results show that the real part of scattering fields for E_x and E_z components has a larger amplitude compared to the imaginary part of the same E_x and E_z. The component of electric field E_y is equal to zero for both real and imaginary cases.

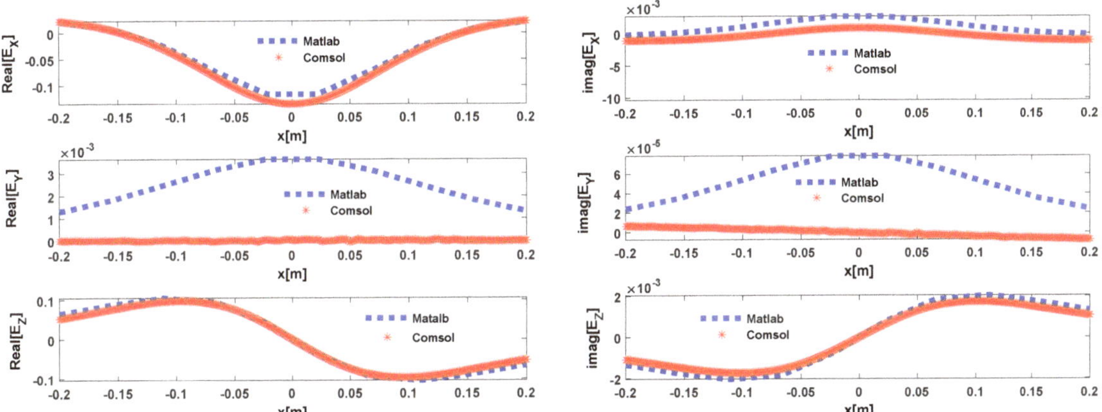

Figure 3. Real part (**left**) and imaginary part (**right**) of the scattered electric field by two PEC spheres. In particular, Matlab results (blue color dot lines) are perfectly superimposed on Comsol results (red marker color lines). Sphere I and sphere II are translated at 0.3 m and -0.3 m along z-axis, and the selected frequency is $f = 3$ MHz.

Figure 4 presents the translation of both PEC spheres I, II translated along z-axis at $z = 0.5$ m and $z = -0.5$ m. The position of the field measuring line is translated at hight $z = 0.2$ m. The observed numerical results at a lower values of truncation sum at $N = 5$ Matlab code show the perfect matching with Comsol results.

3.3. Case 2: f = 0.3 GHz

The numerical results of scattered electric fields have been inspected at a frequency $f = 0.3$ GHz. The PEC spheres I, II have translated along z-axis at $z = 0.5$ m and $z = -0.5$ m. The position of the field measuring line is translated along z-axis at $z = 0.3$ m. During our numerical analysis, the chosen parameters were: radius of spheres I, II $a = c = 0.1$ m, angles of incident $\theta_i = 0$ rad, and $\phi_i = 1$ rad. The best comparison between Matlab code and Comsol simulation results are illustrated in Figure 5.

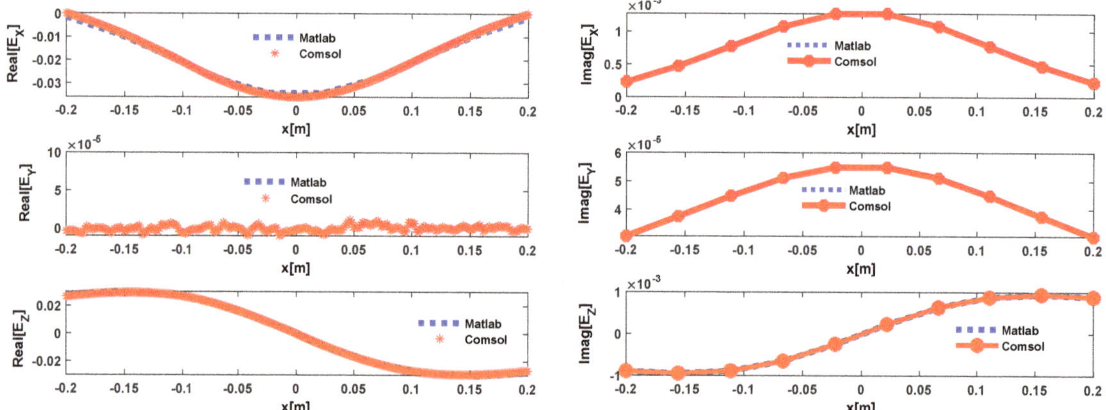

Figure 4. Real part (**left**) and imaginary part (**right**) of the scattered electric field by two PEC spheres. In particular, Matlab results (blue color dot lines) are perfectly super-imposed on Comsol results (red marker color lines). Sphere I and sphere II are translated at 0.5 m and −0.5 m along z-axis, and the selected frequency is $f = 3$ MHz.

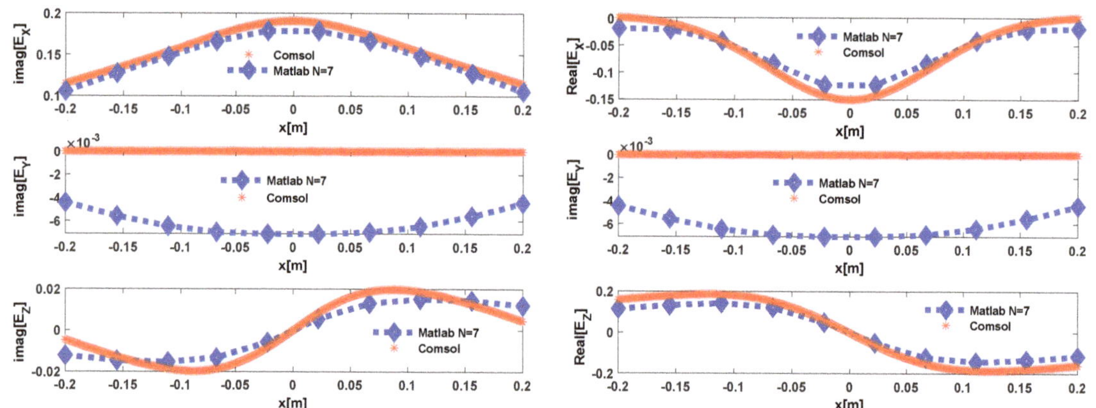

Figure 5. Real part (**left**) and imaginary part (**right**) of the scattered electric field by two PEC spheres. In particular, Matlab results (blue color dot lines) are perfectly super-imposed on Comsol results (red marker color lines). Sphere I and sphere II are translated at 0.5 m and −0.5 m along z-axis, and the selected frequency is $f = 0.3$ GHz.

The numerical results show that the real part of scattering fields for E_z components has a higher amplitude compared to the imaginary part of the same E_z, while the real part of scattering fields for E_x components has a smaller amplitude compared to the imaginary part of the same E_x. The component of electric fields E_y is equal to zero for both real and imaginary cases.

Figure 6 demonstrates the translation of both PEC spheres I, II along z-axis at $z = 0.4$ m and $z = −0.4$ m. The position of the field measuring line is translated at a height $z = 0.1$ m. The numerical results, for a lower value of truncation sum $N = 4$ Matlab code, show fluctuations beyond the results obtained through Comsol. Increasing the truncation sum to $N = 37$ Matlab code makes the numerical results perfectly match with Comsol.

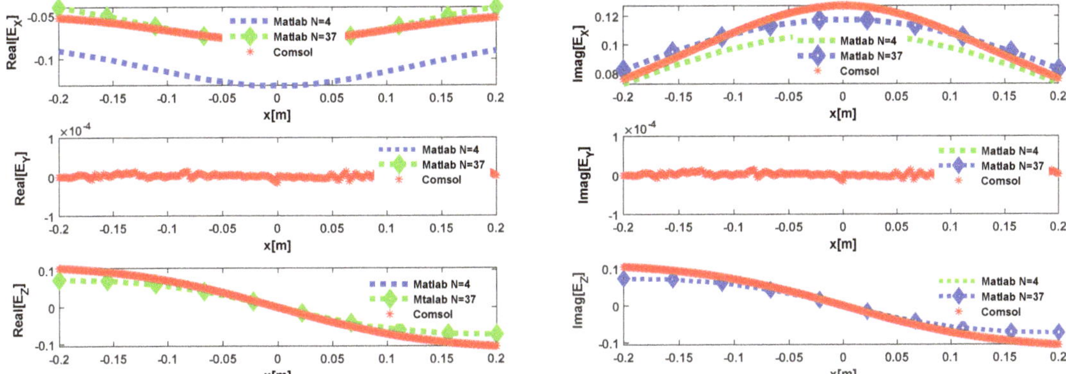

Figure 6. Real part (**left**) and imaginary part (**right**) of the scattered electric field by two PEC spheres. The red color lines (Comsol results), blue and green lines (Matlab results) at different value of the truncation sum N, sphere I, and sphere II are translated at 0.4 m and -0.4 m along z-axis, and the selected frequency is $f = 0.3$ GHz.

3.4. Case 3: f = 3 GHz

The PEC spheres I, II were translated along z-axis at $z = 0.08$ m and $z = -0.08$ m. The position of the field measuring line is translated along z-axis at $z = 0.3$ m. During our numerical test, the following parameters have been chosen: radius of the spheres I, II $a = c = 0.05$ m, angles of incident $\theta_i = 0$ rad, $\phi_i = 1$ rad, and $f = 3$ GHz. Figure 7 illustrates the numerical results for a lower value of truncation sum $N = 5$ Matlab code, showing fluctuations across the Comsol results. Increasing the truncation sum to $N = 40$ in Matlab code makes the numerical results perfectly match with Comsol outcomes.

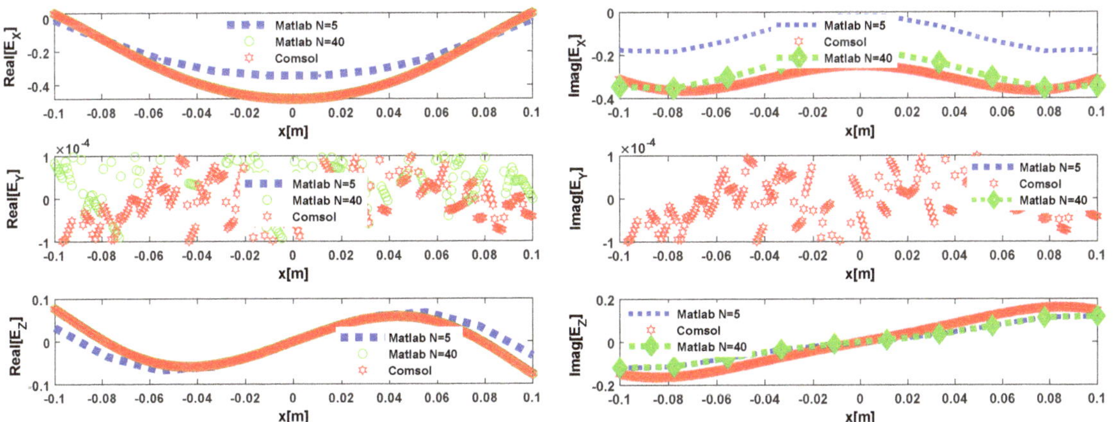

Figure 7. Real part (**left**) and imaginary part (**right**) of the scattered electric field by two PEC spheres. The red color lines (Comsol results), blue and green lines (Matlab results) at different value of the truncation sum N, sphere I and sphere II are translated at 0.08 m and -0.08 m along z-axis, and the selected frequency is $f = 3$ GHz.

4. Truncation Error Analysis of Scattered Electric Field Using Vector Translation AT

The criteria of truncating the inherent infinite series with a finite version of the same to achieve convergence was developed. This leads to a little fluctuation compared to the test of the scattered electric field by two PEC spheres using Comsol and Matlab simulation software, as mentioned in Section 3. The error rate for the proposed approach due to the value of truncation sum N is investigated in this section. Using Equation (49), the error function for scattered electric field is defined as follows:

$$err = \frac{\|E_s^{n+1} - E_s^n\|}{\|E_s^n\|} \qquad n = 1, 2, 3 \ldots \tag{52}$$

where

$$\|E_s^n\| = \sqrt{\sum_{n=1}^{N} |E_s^n|^2} \tag{53}$$

The higher value of truncation number N in the scattered electric field can be employed to tune the accuracy of the solution. Additionally, as the lower values of truncation sum N the error level increases with the variation of frequency. When we increase the number of terms to $N \geq 40$ by using above input, the convergence results are depicted in Figure 8. Finally, the criteria of truncation behavior related to translation AT reliant directly on frequency was discovered. For higher frequency, the truncation sum N converges at a higher numerical value for translation AT shown in Table 1. This methodology is useful even for spheres as small $r << \lambda$ and spheres as large $r >> \lambda$. The former solution is very general and applicable to multiple PEC spheres. When sphere I is moving away from sphere II, for both cases real and imaginary numerical results, a higher concentration of scattering electric field pattern is shown. The improvement of our approach will provide a new concept to researchers for developing the deep analysis of the light scattering problem across general distributions of spheres.

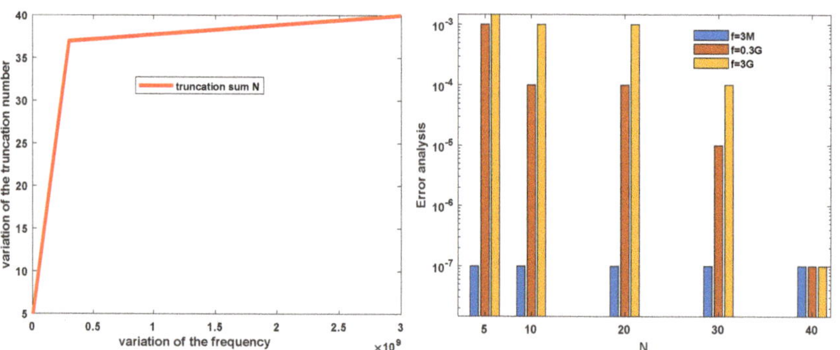

Figure 8. Representation of the truncation variation sum N (**left**) and its convergence solution with the variation of different frequency $f = 3\,\text{M}$, $f = 0.3\,\text{G}$, $f = 3\,\text{G}$ (**right**).

Table 1. Variation of truncation number N with respect to frequency.

#	Frequency	Size of a Sphere	Convergence of the Solution (Truncation Number N)
1	3 MHz	0.1 m	N = 5
1	0.3 GHz	0.1 m	N = 37
2	3 GHz	0.05 m	N = 40

5. Conclusions

This paper presents and explores multiple scattering by two PEC spheres through computer simulations. A method of translation AT for spherical vector wave equation was introduced. To describe this problem, a Cruzan's formalism for vector translation coefficients was proposed. The scattering electric field by two PEC spheres in a near zone region was calculated. The translation of spheres I, II along the z-axis was thought out and resolved to understand the scattering behavior. The results of our calculations were investigated using a numerical approach. The numerical results were compared with the Comsol simulation models and both software results were found to be perfectly matched one another, which gives a fairly good validation.

In summary, the work presented in this letter has investigated the finite truncation sum N. The convergence of the solution is useful to determine the scattering efficiency of a homogeneous spherical scatterer by a plane wave. From a practical point of view, the proposed mathematical expressions are clearly useful to facilitate fast simulations of light scattering phenomena. It should be noted that this truncation criterion applies to homogeneous spheres, whereas other shapes result in their own scattering coefficients.

We aim to build on shared fundamentals, highlight the most pressing research challenges, and exchange state-of-the-art methodologies and approaches. This all-community approach will promote solutions to those issues facing many practical applications in complex media through the rigorous underpinning of mathematical techniques and the development of effective computational methods.

Author Contributions: The authors S.B. conceived the presented idea. S.B. developed the formulation or evolution of overarching research goals and aims. and performed the computations. Following co-authors (M.N., L.D., F.M., F.F.) encouraged first author to investigate a specific aspect and supervised the findings of this work. All authors provided critical feedback and helped in equal contribution to the formation of the research article. All authors have read and agreed to the published version of the manuscript.

Funding: This research received external funding from Sapienza university of Rome Italy.

Data Availability Statement: The authors confirm that the data supporting the findings of this study are available within the article.

Conflicts of Interest: The authors declare no conflict of interest.

References

1. Oguchi, T. Electromagnetic wave propagation and scattering in rain and other hydrometeor. *Proc. IEEE* **1983**, *71*, 1029–1078. [CrossRef]
2. Batool, S.; Dinia, L.; Frezza, F.; Mangini, F.; Nisar, M. Electromagnetic interaction with a monodispersed system in sedimentation equilibrium. In Proceedings of the 2020 XXXIIIrd General Assembly and Scientific Symposium of the International Union of Radio Science, Rome, Italy, 29 August–5 September 2020; pp. 1–4.
3. Borghese, F.; Denti, P.; Saija, R.; Toscano, G.; Sindoni, O.I. Use of group theory for the description of electromagnetic scattering from molecular systems. *J. Opt. Soc. Am. A* **1984**, *2*, 183–191. [CrossRef]
4. Mangini, F.; Dinia, L.; Frezza, F. Electromagnetic scattering by a cylinder in a lossy medium of an inhomogeneous elliptically polarized plane wave. *J. Telecommun. Inf. Technol.* **2019**, *4*, 36–42. [CrossRef]
5. Dinia, L.; Mangini, F.; Frezza, F. Electromagnetic scattering of inhomogeneous plane wave by ensemble of cylinders. *J. Telecommun. Inf. Technol.* **2020**, *3*, 86–92. [CrossRef]
6. Mangini, F.; Tedeschi, N. Scattering of an electromagnetic plane wave by a sphere embedded in a cylinder. *J. Opt. Soc. Am. A* **2017**, *34*, 760–769. [CrossRef]
7. Batool, S.; Naqvi, Q.A.; Fiaz, M.A. Scattering from a cylindrical obstacle deeply buried beneath a planar non-integer dimensional dielectric slab using Kobayashi potential method. *Opt.-Int. J. Light Electron. Opt.* **2018**, *153*, 95–108. [CrossRef]
8. Batool, S.; Nisar, M.; Mangini, F.; Frezza, F.; Fazio, E. Polarization Imaging for Identifying the Microscopical Orientation of Biological Structures. In Proceedings of the URSI GASS Conference, Rome, Italy, 29 August–5 September 2020.
9. Fazio, E.; Batool, S.; Nisar, M.; Mangini, F.; Frezza, F. Recognition of bio-structural anisotropy by polarization-resolved imaging. *Sensor* 2021, *submitted*.
10. Batool, S.; Nisar, M.; Mangini, F.; Frezza, F.; Fazio, E. Scattering of Light from the Systemic Circulatory System. *Diagnostics* **2020**, *10*, 1026. [CrossRef] [PubMed]

11. Batool, S.; Frezza, F.; Mangini, F.; Simeoni, P. Introduction of Radar Scattering Application in Remote Sensing and Diagnostics. *Atmosphere* **2020**, *11*, 517. [CrossRef]
12. Gastellu, E.J.P.; Demarez, V.; Pinel, V.; Zagolski, F. Modeling radiative transfer in heterogeneous 3-D vegetation canopies. *Remote Sens. Environ.* **1996**, *58*, 131–156. [CrossRef]
13. Ioannidou, M.P.; Skaropoulos, N.C. Study of interactive scattering by clusters of spheres. *JOSA A* **1995**, *12*, 1782–1789. [CrossRef]
14. Liang, C.; Lo, Y.T. Scattering by two spheres. *Radio Sci.* **1967**, *12*, 1481–1495. [CrossRef]
15. Olaofe, G.O. Scattering by two cylinders. *Radio Sci.* **1970**, *5*, 351–360. [CrossRef]
16. Fuller, K.A. Optical resonances and two-sphere systems. *Appl. Opt.* **1991**, *30*, 4716–4731. [CrossRef]
17. Bruning, J.; Lo, Y. Multiple scattering of EM waves by spheres part I Multipole expansion and ray-optical solutions. *IEEE Trans. Antennas Propag.* **1971**, *19*, 378–390. [CrossRef]
18. Wang, Y.M.; Chew, W.C. A recursive T-matrix approach for the solution of electromagnetic scattering by many spheres. *IEEE Trans. Antennas Propag.* **1993**, *41*, 1633–1690. [CrossRef]
19. Friedman, B.; Russek, J. Addition theorems for spherical waves. *Q. Appl. Math.* **1954**, *1*, 13–23. [CrossRef]
20. Stein, S. Addition theorems for spherical wave functions. *Q. Appl. Math.* **1961**, *19*, 15–24. [CrossRef]
21. Cruzan, O.R. Translational addition theorems for spherical vector wave functions. *Q. Appl. Math.* **1962**, *20*, 33–40. [CrossRef]
22. Mackowski, D.W. Analysis of radiative scattering for multiple sphere configurations. *Proc. R. Soc. London. Ser. A Math. Phys. Sci.* **1991**, *433*, 599–614.
23. Xu, Y.L. Electromagnetic scattering by an aggregate of spheres. *Appl. Opt.* **1995**, *34*, 4573–4588. [CrossRef]
24. Xu, Y.L. Fast evaluation of the Gaunt coefficients. *Math. Comput.* **1996**, *65*, 1601–1612. [CrossRef]
25. Xu, Y.L. Efficient evaluation of vector translation coefficients in multiparticle light-scattering theories. *J. Comput. Phys.* **1998**, *139*, 137–165. [CrossRef]
26. Batool, S.; Frezza, F.; Mangini, F.; Xu, Y.L. Scattering from multiple PEC sphere using translation addition theorems for spherical vector wave function. *J. Quant. Spectrosc. Radiat. Transf.* **2020**, *248*, 106905. [CrossRef]
27. Batool, S.; Benodetti, A.; Frezza, F.; Mangini, F.; Xu, Y.L. Effect of finite terms on the truncation error of addition theorems for spherical vector wave function. In Proceedings of the Photonics and Electromagnetics Research Symposium-Spring (PIERS-Spring), Rome, Italy, 17–20 June 2019; pp. 2795–2801.
28. Neves, A.A.; Pisignano, D. Effect of finite terms on the truncation error of Mie series. *J. Comput. Phys.* **2012**, *37*, 2418–2420.
29. Wiscombe, W.J. Improved Mie scattering algorithms. *Appl. Opt.* **1980**, *19*, 1505–1509. [CrossRef]
30. Frezza, F.; Mangini, F.; Tedeschi, N. Introduction to electromagnetic scattering: Tutorial. *J. Opt. Soc. Am.* **2018**, *35*, 163–173. [CrossRef] [PubMed]
31. Frezza, F.; Mangini, F.; Tedeschi, N. Introduction to electromagnetic scattering, part II: Tutorial. *J. Opt. Soc. Am.* **2020**, *37*, 1300–1315. [CrossRef]
32. Mackowski, D.W. Calculation of total cross sections of multiple-sphere clusters. *J. Opt. Soc. Am. A* **1994**, *11*, 2851–2861. [CrossRef]

Article

Recognition of Bio-Structural Anisotropy by Polarization Resolved Imaging

Eugenio Fazio [1,*], Sidra Batool [1,2], Mehwish Nisar [1,2], Massimo Alonzo [3] and Fabrizio Frezza [2]

[1] Dipartimento di Scienze di Base e Applicate per l'Ingegneria, Sapienza Università di Roma, 00161 Roma, Italy; sidra.batool@uniroma1.it (S.B.); mehwish.nisar@uniroma1.it (M.N.)

[2] Dipartimento di Ingegneria dell'Informazione Elettronica e Telecomunicazioni, Sapienza Università di Roma, 00184 Roma, Italy; fabrizio.frezza@uniroma1.it

[3] Fusion and Nuclear Safety Department, ENEA—Casaccia Research Center, 00123 Roma, Italy; massimo.alonzo@enea.it

* Correspondence: eugenio.fazio@uniroma1.it

Abstract: In this paper, we develop a simple technique to identify material texture from far, by using polarization-resolved imaging. Such a technique can be easily implemented into industrial environments, where fast and cheap sensors are required. The technique has been applied to both isotropic references (Teflon bar) and anisotropic samples (wood). By studying the radiance of the samples illuminated by linearly polarized light, different and specific behaviours are identified for both isotropic and anisotropic samples, in terms of multipolar emission and linear dichroism, from which fibre orientation can be resolved.

Keywords: light polarization; polarimetry; polarized imaging; augmented reality; industrial sensing

Citation: Fazio, E.; Batool, S.; Nisar, M.; Alonzo, M.; Frezza, F. Recognition of Bio-Structural Anisotropy by Polarization Resolved Imaging. *Electronics* **2022**, *11*, 255. https://doi.org/10.3390/electronics11020255

Academic Editor: Reza K. Amineh

Received: 29 November 2021
Accepted: 11 January 2022
Published: 13 January 2022

Publisher's Note: MDPI stays neutral with regard to jurisdictional claims in published maps and institutional affiliations.

1. Introduction

Most natural matter is composite, heterogeneous, made up of different materials, present in several phases, and with properties and characteristics different from their bulk equivalents. A composite is presented as a set of micro and nanostructures assembled together: for example, the tissues of living beings, both animal and vegetable, are formed by cells, fibres, membranes, layers, i.e., they are not made up of homogeneous substances but by a set of small structures, aggregated together. Just think of the muscle or wood tissue, both made up of fibres connected together that make it resistant and elastic. Technology has taken an example from nature, engineering materials through the creation of composites in order to create new ones with properties not found in nature. For example, you can think about materials with glass or carbon fibres incorporated in resins or sheathed by adhesive plastic films, in order to create bulks or tissues that are both light and extremely resistant at the same time.

To visualize and characterize the micro- and nano-structuring of a composite, optical microscopy techniques are used, which can be both simple and advanced, such as fluorescence [1] or nonlinear [2]. Microscopy directly observes the individual fibres (even the etymology suggests a "vision at the micron scale") and, through appropriate image processing, characterizes the order of structuring of the samples studied. The propagation of light in biological tissues [3,4] has always attracted a lot of attention due to the infinite implications that it can entail. For example, consider the possibility of recognizing diseases [5–8] or discriminating similar tissues [9,10]. To study anisotropic materials, the polarization of light is often used as a probe [4–18]: by using electromagnetic fields oscillating parallel or orthogonally to the dipoles of the material, it is also possible to obtain an anisotropic diffusion according to the microscopic alignment of the material. Thus, through the analysis of microscopic images taken using polarized light, it is possible to characterize the morphology of different materials [11,16,17].

By microellissometry [18], or through the study of Mueller matrices [19–25], it is possible to measure the state of polarization of the light and obtain information on the microscopic or nanoscopic order of the material. In fact, it has been widely documented that both the propagation of light in nanostructured materials [26–29] and the Lambertian-type emission from non-symmetrical nano-oscillators [30–34] maintain information on the orientation and order of the material.

However, optical microscopy requires time and specific preparation of the samples, which could result in handicaps in the industrial field where, otherwise, fast, economical techniques are required and that can work at great distances, even without special preparation of the samples.

The fourth industrial revolution we are experiencing, known as Industry 4.0, sees a propensity of today's automation to insert new production procedures, aimed at improving working conditions, creating new business models but above all at increasing the yield of the plants by improving quality. of products. The evolution of industrial computer applications passes through the use of innovative vision and recognition techniques, such as machine learning and augmented reality. These require the support of new sensory techniques, that is, the use of innovative electromagnetic measuring instruments, capable of providing rapid and precise information remotely.

Thus, we asked ourselves whether it is possible to extract information from polarization-resolved images shot at a distance with a simple camera or a video camera, in order to discriminate and characterize the structuring of materials. For this purpose, we acquired and processed images with polarization parallel and orthogonal to that of illumination, highlighting how these are affected by the structuring of the material and its orientation in space. In fact, by studying the polar distribution of the emission in terms of multipolar contributions, up to the fourth order, it is possible to observe that isotropic materials as well as anisotropic ones, with specific spatial orientations, show different optical behaviours.

Multipole analysis of light emission and scattering [35–37] has been effectively used to study micro- and nanospheroid [38–40], acoustic scattering [41], or light scattered from single biological cells [42]. This is an alternative procedure to the representation of Mueller matrices [24,25], from which to immediately identify whether the material is isotropic or anisotropic and possibly also its spatial orientation.

The present work studies two different types of material: Teflon, as an isotropic reference, and wood, as an anisotropic material. Wood is a good material for optical investigations, being formed by micro- and nano-fibrils [43] oriented along a privileged direction; wood has already been studied by optical techniques in the past to identify structural changes induced by mechanical stresses [44].

2. Materials and Methods

The experimental setup is schematically shown in Figure 1. A sample is homogeneously illuminated from the orthogonal direction (azimuthal angle = 0, polar angle = $\pi/2$) by a collimated beam of linearly-polarized light, generated by a blue/violet LED.

The light color is chosen because the short wavelength (405 nm) gives higher sensitivity in resolving texture nano-orientations.

Using an achromatic zoom (180×), a CCD camera (whose response linearity has been previously verified) records the images of the sample at a great distance (about 2 m) and at different azimuthal viewing angles θ_{view} (variable from 10° to 70° with 10° steps, at fixed polar angle $\theta_{polar} = \pi/2$). At each viewing angle, the linear-polarization direction of the illumination is rotated all over 360° (in steps of 10°) and, for each orientation, two images are taken of the parallel and orthogonal polarised light respectively (with respect to the input one). As a reference, the 0° polarisation means TE on the sample (as well as 180° and 360°), while 90° means TM (as well as 270°).

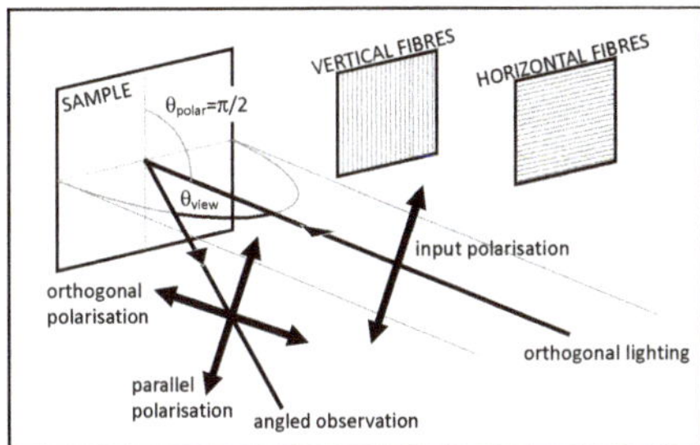

Figure 1. Experimental geometry. A sample is orthogonally illuminated by incoherent linearly-polarized light. A camera records images of the sample surface at different azimuthal angles θ_{view}, performing a selection of the observed light polarizations, both parallel and orthogonal to the input one.

From the digitalised images, the average radiance is calculated for each viewing angle over homogeneous areas of 200×200 pixels which correspond to roughly 2 mm² on the samples (with a signal depth of 8 bits). This particular geometry (average on a fixed area of the image) was chosen to compensate for the angular dispersion of Lambertian emitters. The emissivity of a Lambertian emitter is maximum in the orthogonal direction ($\theta_{polar} = \pi/2$, $\theta_{view} = 0$) and decreases moving away angularly from it, with a cosine type trend. To compensate for this dependence, fixed image areas have been selected which correspond to increasingly larger surfaces on the sample if θ_{view} is increased, with a $1/\text{cosine}$ trend. The reciprocal compensation leads to a "clean" observation of the emissivity, formally independent of the angle. Figure 2 shows the average emissivity of a Teflon surface (isotropic reference) illuminated with both TE ($0°$) and TM ($90°$) polarisations. As you can see, the reported signals are not affected by the typical Lambertian angular dependence.

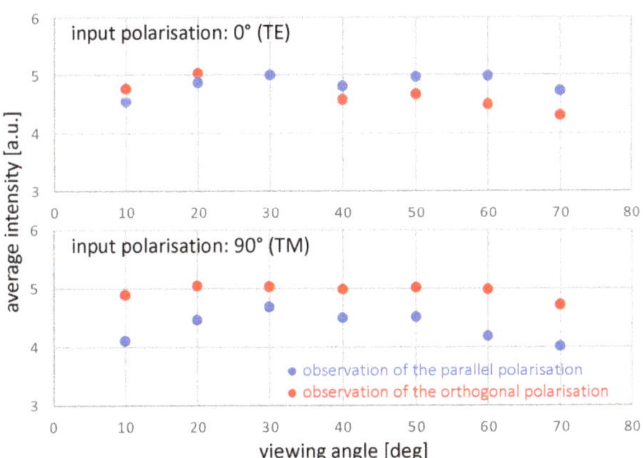

Figure 2. Observed average intensity of the Teflon radiance illuminating with TE ($0°$) or TM polarizations ($90°$).

In order to have a good comparison between isotropic and structured materials, an opaque white Teflon tablet and a wooden one were chosen respectively (Figure 3). The latter is made of pine wood, obtained from the xylem with a radial cut. In this way, fibres and fibrils constituting the wooden fabric are oriented parallel with respect to the surface. According to the geometry shown in Figure 1, the wood sample was positioned with either vertical (parallel to the polar direction) or horizontal fibres (i.e., in the equatorial plane). As a consequence, vertical fibres were parallel to the electric field for the TE polarization ($0°$), while horizontal fibres were orthogonal to the electric field for the TM polarization ($90°$).

Figure 3. Wood and Teflon samples. The wood one shows the micro-structure of macroscopic fibres.

The polarisation-dependent emissivity diagrams have been numerically processed in order to identify the origin of the signals and to discriminate those most sensitive to the texture orientation. More specifically, the polarization-dependent trends have been numerically processed by means of Fourier series fitting. For this purpose, the MatLab "Curve Fitting Tool" package was used. Among all the infinite terms, we have used the first four which correspond to the monopole (zero-order/constant term), dipole (first-order), quadrupole (second-order), and octupole (fourth-order) emitters. The third term has been fixed to zero during the fitting procedure because it has no physical meaning. The frequency of the Fourier series was fixed at $2\pi/360 = 0.01745$ rad/$^\circ$.

3. Results

In Figure 4, the polar distributions of the radiance emitted by the Teflon sample are reported to vary the polarization plane of the lightning light. Each figure was recorded at a different angle of observation, ranging from 10° to 70°. The red curves in the figure describe the radiant intensity with polarization parallel to the illumination, while the blue ones represent the radiant intensity with orthogonal polarization. As you can see, towards the orthogonal observation (see 10°) the two curves tend to achieve similar circular shapes (which should coincide for observation at 0° where the TE and TM polarisations coincide). The orthogonal polarization (blue line) has a more flattened trend: this flattening increases with the angle. The red curve is also flattened more by increasing the angle but in the other direction. As you can be seen, although Teflon is an isotropic diffuser, the specific geometry maintains an anisotropic behaviour due to the different TE and TM emissivities, which depend on the emission angle too.

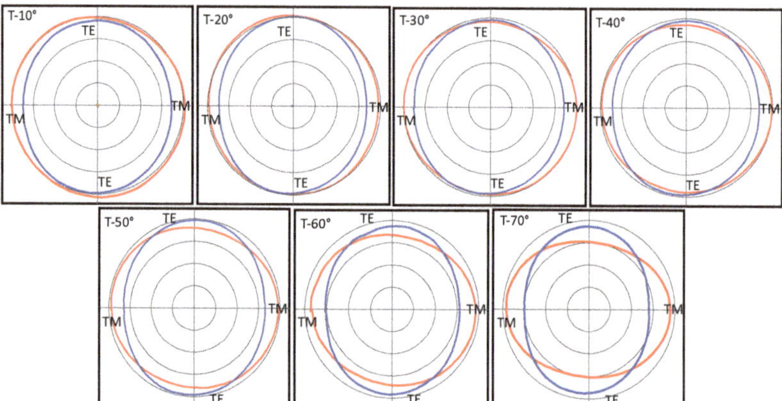

Figure 4. Polar distributions of Teflon radiances for polarizations parallel (red) and orthogonal (blue) to the input one. Images report the measured radiances varying the observation azimuthal angle θ_{view}.

The equivalent radiance of the unisotropic wood samples is shown in Figure 5 for the two orientations of wood with vertical (W-VF) and horizontal (W-HF) fibres. One more time, W-VF means fibres parallel to the TE polarisation while W-HF means fibres parallel to the TM one.

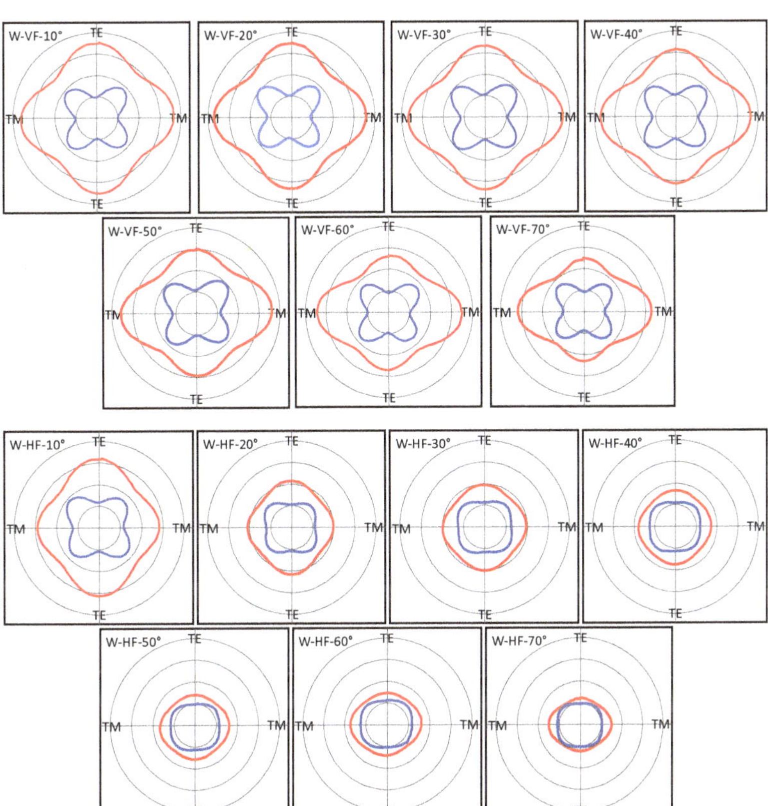

Figure 5. Polar distributions of wood radiances for polarizations parallel (red) and orthogonal (blue) to the input one. Images report the measured radiances varying the observation azimuthal angle θ_{view}. Top images refer to wood with vertical fibres (W-VF); bottom images refer to wood with horizontal fibres (W-HF).

In this case, the distributions of the parallel and orthogonal radiances are very different from each other and do not tend to assume the same distribution going towards the orthogonal observation.

The parallel radiance of the vertical fibres flattens increasing the viewing angle, similarly to that shown in Figure 4 for the isotropic diffuser. This means that the TE emission decreases with the viewing angle much more than the TM one. Instead, the polar distribution of the radiance from horizontal fibres changes significantly for both TE and TM polarisations: the whole emission efficiency becomes lower and lower as the angle increases, and the shape of the distribution also changes.

The polar distributions at the orthogonal polarisation suffer less the viewing angle: the emission from vertical fibres seems completely insensitive to the viewing angle, while for horizontal fibres the main differences appear at 45°-135°-225°-315°: the "flower" type emission is rapidly attenuated, tending to a "squared" distribution. Such variations clearly describe different weights of the multipolar emissive efficiencies. In order to discriminate each contribution, different Fourier elements of the angular distributions were calculated, corresponding to monopole, dipole, quadrupole and octupole emissions. In Figure 6 the normalized multipole contributions are reported. The values related to polarization parallel to the illumination are shown in red while the orthogonal ones are shown in blue.

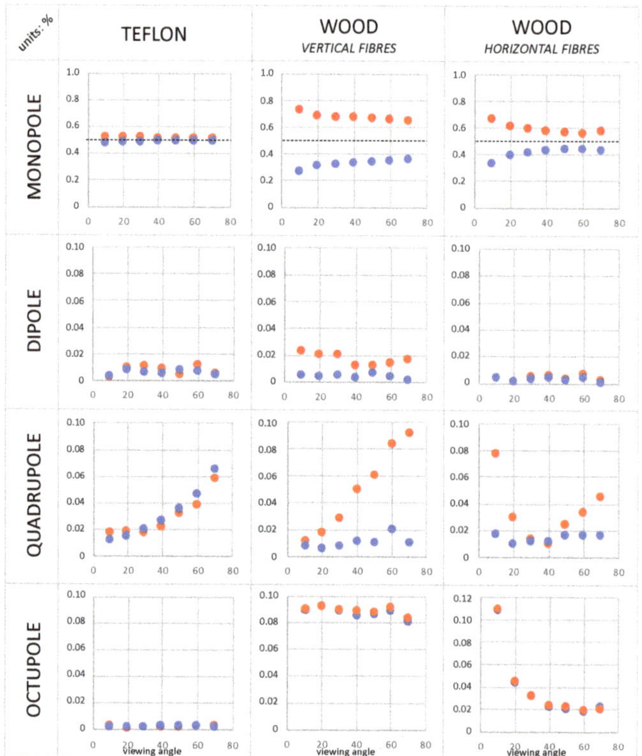

Figure 6. Multi-pole analysis of the obtained polar signals shown in Figures 4 and 5.

The dominant contributions are due to monopole emission, as expectable. The Teflon emission is almost completely depolarized (both signals are close to 50%), with a very slight memory of the lightning polarization (red signals). Such memory is attenuated by increasing the viewing angle, towards a perfect depolarization by tending to 90°. This behaviour occurs because Teflon is a "soft" diffuser, and consequently the radiation penetrates inside it before coming out backward, maintaining information of the input. In wood, anisotropy is indeed more evident. The information of the initial polarization is maintained. Wood is certainly a stiffer diffuser than Teflon, where light penetration is very limited if not completely absent. For both fibre alignments, the depolarization of the monopole emission decreases by increasing the viewing angle but never disappears.

Dipole contributions for all three samples are very limited to a few percent, almost negligible.

Quadrupole contributions of Teflon are important and grow by increasing the viewing angle, up to almost 10% of the entire signal such phenomena are also present in the quadrupole emission of wood with vertical fibres only in the parallel polarization component; the orthogonal polarization component, on the other hand, is little affected (for about 1%) by the quadrupole emission. This insensitivity is also present for wood with horizontal fibres. In this case, however, the parallel polarization component exhibits an anomalous behaviour, with a minimum in correspondence of about 40° which can be attributed to Brewster-like loss of emissive efficiency.

The octupole contribution is null for the isotropic sample (Teflon) while in wood it shows high values. The vertical orientation of the fibres gives a contribution of octupole independent of the viewing angle while, for the horizontal orientation of the fibres, the emission efficiency of the octupole terms decreases as the angle of observation increases.

Monopole, quadrupole and octupole radiance contributions are indeed discriminating factors to recognize the orientation of the material texture. Monopole and quadrupole emissions are polarisation sensitive while octupole seems to be perfectly depolarised.

4. Discussion

Figure 6 shows which multipole emission contributions can recognise a structural anisotropy of surfaces. The most efficient emission is undoubtedly the monopole one, as expectable: the first element of a power expansion is always more effective than the following higher-order terms. The isotropic Teflon sample shows a monopole anisotropy quantifiable in a 6% difference between the signals emitted on the two polarisations. This effect is due to a different emissivity of a "soft" material, i.e., a surface that does not reflect specularly but whose incident radiation penetrates and suffers multiple scattering in the volume before being re-emitted backward. The phenomenon is well described by the Kubelka–Munk theory [45,46]. At larger viewing angles, the number of multi-scattering events strongly increases, inducing more efficient depolarisation.

Unlike the isotropic sample, wood shows a stronger monopole anisotropic radiance, quantifiable for a normal observation at about 50–60% of the total signal. Vertical fibres retain more information than horizontal fibres by means of a geometrical shadowing effect. As the observation angle varies, vertical fibres pack up in an increasingly dense manner, showing a stronger anisotropic emissivity (Figure 7).

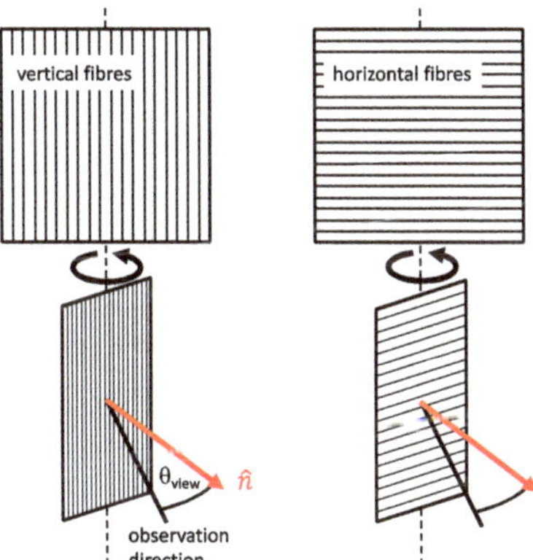

Figure 7. Wood observations at different azimuthal angles θ_{view} are influenced by the fibre orientation.

Instead, the package shadowing does not affect the horizontal fibres, that remain equally spaced, without visibility alteration. This behaviour is well known and experimented with by everyone observing for example the grass of a stadium that appears striped depending on how it has been cut/combed.

Dipole contributions are extremely low (of the order of 1%), and therefore, negligible.

Instead, the quadrupole contributions are important and show different trends for the parallel and orthogonal radiative polarizations. Teflon shows very similar trends on both crossed polarizations. Different behaviours occur for wood, where the crossed emission (i.e., orthogonal to the lightning polarization) is always negligible, reaching a value between 1 and 2%. We do expect that higher polar emissivity is related to smaller textures.

Otherwise, the parallel polarization emission of vertical fibres becomes more and more efficient with the viewing angle. This would be the behaviour for both vertical and horizontal fibres, were it not for the fact that horizontal fibres show a Brewster-like emission quenching at about 30–40°.

Octupole terms are important to recognise wood texture but do not show any anisotropic behaviour.

The quadrupolar anisotropic behaviour can be highlighted by means of a linear dichroism *LD* factor, defined as the difference between the parallel and orthogonal radiances normalized to their sum:

$$LD = \frac{R_{\parallel} - R_{\perp}}{R_{\parallel} + R_{\perp}} \tag{1}$$

The *LD* trends are shown in Figure 8. Teflon dichroism (upper part of Figure 8) is indeed very small, lower than 10%: it has been magnified in order to resolve the slight modifications. A small viewing angles dichroism shows a dipolar trend; at higher angles the quadrupole contribution becomes more and more. This anisotropy is mainly geometric in nature, recognising TE and TM polarisation emissions.

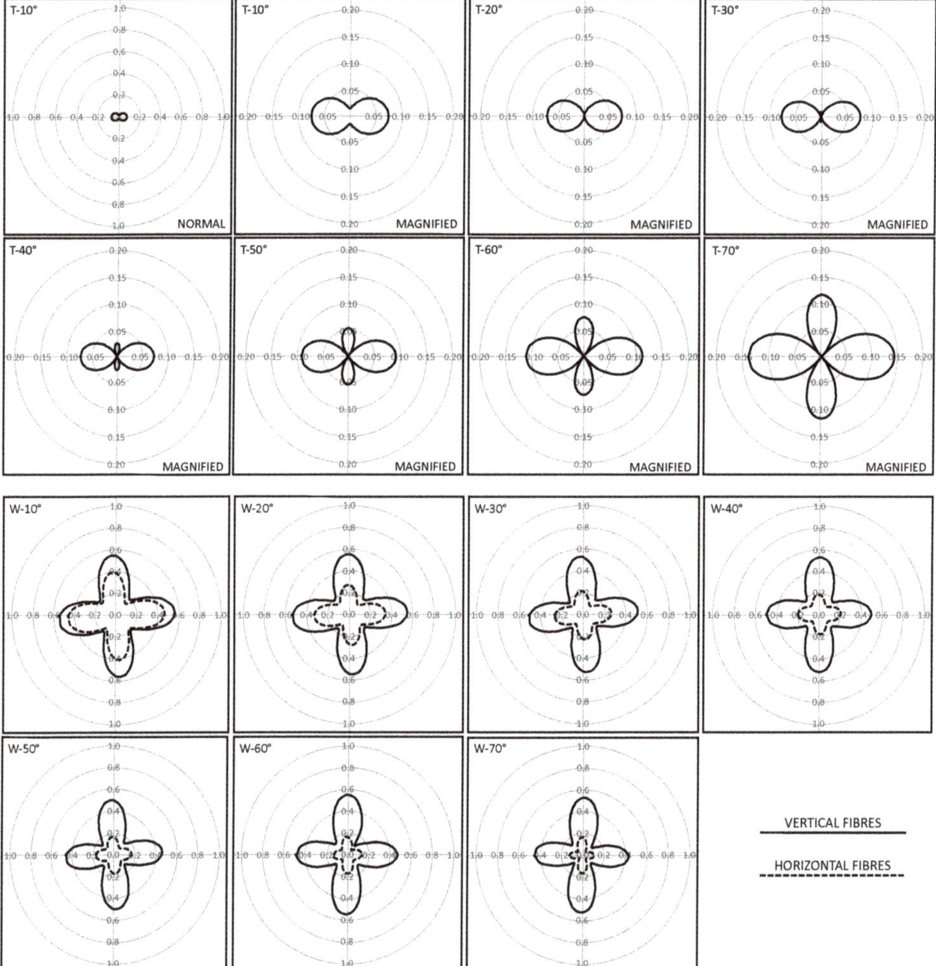

Figure 8. Polar representation of linear dichroism as the azimuth angle of observation varies, both for the Teflon samples (8 upper panels) and for those of wood (8 lower panels). The wood representations report both the experimental results for vertically (solid line) and horizontally (dashed line) oriented fibres.

Wood dichroism (lower part of Figure 8) is indeed more effective, arriving at values up to 0.6. *LD*'s deriving for vertical (solid line) and horizontal (dashed line) fibres are reported on the same graphs for comparison.

As you can see, the linear dichroism of vertical fibres is much more intense than that of horizontal fibres and remains high almost insensitive to the observation angle. On the other hand, the horizontal fibres are strongly affected by the Lambertian behaviour and dichroism tends to quench for viewing angles close to 90°. In both cases, vertical fibres and horizontal fibres, the shape of the dichroism is clearly linked to the quadrupole emission.

5. Conclusions

In conclusion, the performed work has shown that through polarization-resolved imaging it is possible to discriminate between isotropic and anisotropic materials, but also to recognise different texture orientations.

Different textures are discriminated monitoring monopole, quadrupole and octupole contributions to emissions. The discriminator par excellence of the material's anisotropy is given by the octupole emission, zero for the isotropic reference and different from zero for anisotropic samples. Wood monopole and quadrupole contributions show different behaviours of the polarisation components. Quadrupole crossed radiance (with respect to the lightning polarisation) is always very low; instead, parallel radiance is higher and depends on the observation angle. Orthogonal fibres show a Brewster-like quenching of the quadrupole parallel emission at about 30–40° which is not present with vertical fibres and that can be effectively used as a fibre-orientation detector. We do expect that higher-order polar emissivity would depend on smaller and smaller material texture, but this point is at the moment under investigation.

Author Contributions: Conceptualization, E.F.; methodology, E.F.; software, M.A.; validation, E.F., S.B., M.N.; formal analysis, E.F.; investigation, S.B., M.N.; data curation, E.F.; writing, E.F., M.A., F.F.; writing; supervision, E.F., F.F. All authors have read and agreed to the published version of the manuscript.

Funding: Sapienza Università di Roma, Progetto Ateneo 2107623 Prot. RG120172B9D9AC5A. AMDG.

Data Availability Statement: The experimental data are available on specific request.

Conflicts of Interest: The authors declare no conflict of interest.

References

1. Levitt, J.A.; Matthews, D.R.; Ameer-Beg, S.M.; Suhling, K. Fluorescence lifetime and polarization-resolved imaging in cell biology. *Curr. Opin. Biotechnol.* **2009**, *20*, 28–36. [CrossRef] [PubMed]
2. Goto, A.; Otomo, K.; Nemoto, T. Real-Time Polarization-Resolved Imaging of Living Tissues Based on Two-Photon Excitation Spinning-Disk Confocal Microscopy. *Front. Phys.* **2019**, *56*, 7. [CrossRef]
3. Wilson, B.C.; Adam, G. A Monte Carlo model for the absorption and flux distribution of light in tissue. *Med. Phys.* **1983**, *10*, 824–830. [CrossRef] [PubMed]
4. Maitland, D.J.; Walsh, J.T. Quantitative Measurements of Linear Birefringence During Heating of Native Collagen. *Lasers Surg. Med.* **1997**, *20*, 310–318. [CrossRef]
5. Jacques, S.L.; Lee, K. Polarized video imaging of skin. In Proceedings of the SPIE, San Jose, CA, USA, 1 July 1998; Volume 3245, pp. 356–362.
6. Anastasiadou, M.; De Martino, A.; Clement, D.; Liege, F.; Laude-Boulesteix, B.; Quang, N.; Dreyfuss, J.; Huynh, B.; Schwartz, L.; Cohen, H. Polarimetric imaging for the diagnosis of cervical cancer. *Phys. Stat. Sol.* **2008**, *5*, 1423–1426. [CrossRef]
7. Manhas, S.; Swami, M.S.; Patel, H.S.; Uppal, A.; Ghosh, N.; Gupta, P.K. Polarized diffuse reflectance measurements on cancerous and noncancerous tissues. *J. Biophotonics* **2009**, *2*, 581–587. [CrossRef]
8. Cheong, W.; Prahl, S.A.; Welch, A.J. A review of the Optical Properties of Biological Tissues. *IEEE J. Quant. Electr.* **1990**, *26*, 2166–2185. [CrossRef]
9. Arwin, H. Application of ellipsometry techniques to biological materials. *Thin Solid Films* **2011**, *519*, 2589–2592. [CrossRef]
10. He, C.; He, H.; Chang, J.; Chen, B.; Ma, H.; Booth, M.J. Polarisation optics for biomedical and clinical applications: A review. *Light Sci. Appl.* **2021**, *10*, 194. [CrossRef]
11. Sakai, S.; Nakagawa, N.; Yamanari, M.; Miyazawa, A.; Yasuno, Y.; Matsumoto, M. Relationship between dermal birefringence and the skin surface roughness of photoaged human skin. *J. Biomed. Opt.* **2009**, *14*, 044032. [CrossRef]

12. Ghosh, N.; Vitkin, I.A. Tissue polarimetry: Concepts, challenges, applications and outlook. *J. Biomed. Opt.* **2011**, *16*, 110801. [CrossRef] [PubMed]

13. Chin, L.; Yang, X.; McLaughlin, R.A.; Noble, P.B.; Sampson, D.D. En face parametric imaging of tissue birefringence using polarization-sensitive optical coherence tomography. *J. Biomed. Opt.* **2013**, *18*, 066005. [CrossRef] [PubMed]

14. Tuchin, V.V. Polarized light interaction with tissues. *J. Biomed. Opt.* **2016**, *21*, 071114. [CrossRef] [PubMed]

15. Kienle, A.; Forster, F.K.; Hibst, R. Anisotropy of light propagation in biological tissue. *Opt. Lett.* **2004**, *29*, 2617–2619. [CrossRef] [PubMed]

16. Wallenburg, M.A.; Wood, M.F.G.; Ghosh, N.; Vitkin, I.A. Polarimetry-based method to extract geometry-independent metrics of tissue anisotropy. *Opt. Lett.* **2010**, *35*, 2570–2572. [CrossRef]

17. Laurent, C.; Ahmed, S.; Boardman, R.; Cook, R.; Dyke, G.; Palmer, C.; Schneider, P.; De Kat, R. Imaging techniques for observing laminar geometry in the feather shaft cortex. *J. Microsc.* **2020**, *277*, 154–159. [CrossRef]

18. Cohn, R.F.; Wagner, J.W.; Kruger, J. Dynamic imaging microellipsometry: Theory, system design, and feasibility demonstration. *Appl. Opt.* **1988**, *27*, 4664–4671. [CrossRef]

19. He, H.; Zeng, N.; Li, D.; Liao, R.; Ma, H. Quantitative Mueller matrix polarimetry techniques for biological tissues. *J. Innov. Opt. Health Sci.* **2012**, *5*, 1250017. [CrossRef]

20. Ghosh, N.; Wood, M.F.G.; Vitkin, I.A. Mueller matrix decomposition for extraction of individual polarization parameters from complex turbid media exhibiting multiple scattering, optical activity and linear birefringence. *J. Biomed. Opt.* **2008**, *13*, 044036. [CrossRef]

21. Guo, X.; Wood, M.F.G.; Vitkin, I.A. Angular measurements of light scattered by turbid chiral media using linear Stokes polarimeter. *J. Biomed. Opt.* **2006**, *11*, 041105. [CrossRef]

22. Spandana, K.U.; Mahato, K.K.; Nirmal, M. Polarization-resolved Stokes-Mueller imaging: A review of technology and applications. *Lasers Med. Sci.* **2019**, *34*, 1283–1293.

23. Ghosh, N.; Wood, M.F.G.; Vitkin, I.A. Polarimetry in turbid, birefringent, optically active media: A Monte Carlo study of Mueller matrix decomposition in the backscattering geometry. *J. Appl. Phys.* **2009**, *105*, 102023. [CrossRef]

24. Batool, S.; Nisar, M.; Mangini, F.; Frezza, F.; Fazio, E. Polarization Imaging for Identifying the Microscopical Orientation of Biological Structures. In Proceedings of the URSI GASS 2020 Conference, Rome, Italy, 29 August–5 September 2020.

25. Batool, S.; Nisar, M.; Mangini, F.; Frezza, F.; Fazio, E. To study the Mueller matrix polarimetry for the characterization of wood and Teflon flat samples. *Results Opt.* **2021**, *4*, 100102. [CrossRef]

26. Van Tiggelen, B.A.; Maynard, R.; Heiderich, A. Anisotropic Light Diffusion in Oriented Nematic Liquid Crystals. *Phys. Rev. Lett.* **1996**, *77*, 639–642. [CrossRef] [PubMed]

27. Johnson, P.M.; Bret, B.P.J.; Rivas, J.G.; Kelly, J.J.; Lagendijk, A. Anisotropic Diffusion of Light in a Strongly Scattering Material. *Phys. Rev. Lett.* **2002**, *89*, 243901. [CrossRef] [PubMed]

28. Johnson, P.M.; Faez, S.; Lagendijk, A. Full characterization of anisotropic diffuse light. *Opt. Expr.* **2008**, *16*, 7435–7446. [CrossRef]

29. Alerstam, E.; Svensson, T. Observation of anisotropic diffusion of light in compacted granular porous materials. *Phys. Rev. E* **2012**, *85*, 040301. [CrossRef] [PubMed]

30. Grzela, G. Directional Light Emission and Absorption by Semiconductor Nanowires. Ph.D. Thesis, Technische Universiteit Eindhoven, Eindhoven, The Netherlands, 2013.

31. Lorente, A.; Boersma, K.F.; Stammes, P.; Tilstra, L.G.; Richter, A.; Yu, H.; Kharbouche, S.; Muller, J.P. The importance of surface reflectance anisotropy for cloud and NO2 retrievals from GOME-2 and OMI. *Atmos. Meas. Tech.* **2018**, *11*, 4509–4529. [CrossRef]

32. VanRie, J.; Schütz, C.; Gencer, A.; Lombardo, S.; Gasser, U.; Kumar, S.; Salazar-Alvarez, G.; Kang, K.; Thielemans, W. Anisotropic Diffusion and Phase Behavior of Cellulose Nanocrystal Suspensions. *Langmuir* **2019**, *35*, 2289–2302. [CrossRef] [PubMed]

33. Gallinelli, T.; Barbet, A.; Druon, F.; Balembois, F.; Georges, P.; Billeton, T.; Chenais, S.; Forget, S. Enhancing brightness of Lambertian light sources with luminescent concentrators: The light extraction issue. *Opt. Express* **2019**, *27*, 11830–11843. [CrossRef]

34. Chen, J.; Bhattarai, R.; Cu, J.; Shen, X.; Hoang, T. Anisotropic optical properties of single Si_2Te_3 nanoplates. *Sci. Rep.* **2020**, *10*, 19205. [CrossRef]

35. Poutrina, E.; Urbas, A. Multipole analysis of unidirectional light scattering from plasmonic dimers. *J. Opt.* **2014**, *16*, 114005. [CrossRef]

36. Alaee, R.; Rockstuhl, C.; Fernandez-Corbaton, I. An electromagnetic multipole expansion beyond the long-wavelength approximation. *Opt. Commun.* **2018**, *407*, 17–21. [CrossRef]

37. Evlyukhin, A.B.; Chichkov, B.N. Multipole decompositions for directional light scattering. *Phys. Rev. B* **2019**, *100*, 125415. [CrossRef]

38. Chu, L.; Wang, S. Multipole expansion of the scattering of light from a metal microspheroid. *J. Opt. Soc. Am. B* **1985**, *2*, 950–955. [CrossRef]

39. Evlyukhin, A.B.; Reinhardt, C.; Chichkov, B.N. Multipole light scattering by nonspherical nanoparticles in the discrete dipole approximation. *Phys. Rev. B* **2011**, *23*, 235429. [CrossRef]

40. Evlyukhin, A.B.; Reinhardt, C.; Evlyukhin, E.; Chichkov, B.N. Multipole analysis of light scattering by arbitrary-shaped nanoparticles on a plane surface. *J. Opt. Soc. Am. B* **2013**, *30*, 2589–2598. [CrossRef]

41. Imhof, M.G. Multiple multipole expansions for acoustic scattering. *J. Acoust. Soc. Am.* **1995**, *97*, 754. [CrossRef]

42. Videen, G.; Ngo, D. Light scattering multipole solution for a cell. *J. Biomed. Opt.* **1998**, *3*, 212–220. [CrossRef]
43. Melelli, A.; Jamme, F.; Legland, D.; Beaugrand, J.; Bourmaud, A. Microfibril angle of elementary flax fibres investigated with polarised second harmonic generation microscopy. *Ind. Crops Prod.* **2020**, *156*, 112847. [CrossRef]
44. Niskanen, I.; Räty, J.; Soetedjo, H.; Hibino, K.; Oohashi, H.; Heikkilä, R.; Matsuda, K.; Otani, Y. Measurement of the degree of polarisation of thermally modified Scots pine using a Stokes imaging polarimeter. *Opt. Rev.* **2020**, *27*, 178–182. [CrossRef]
45. Kubelka, P.; Munk, F. Ein Beitrag zur Optik der Farbanstriche. *Z. Tech. Phys.* **1931**, *12*, 593–601.
46. Sandoval, C.; Kim, A.D. Generalized Kubelka-Munk approximation for multiple scattering of polarized light. *J. Opt. Soc. Am. A* **2017**, *34*, 153–160. [CrossRef] [PubMed]

Article

Characterizing THz Scattering Loss in Nano-Scale SOI Waveguides Exhibiting Stochastic Surface Roughness with Exponential Autocorrelation

Brian Guiana [†] and Ata Zadehgol *,[†]

Department of Electrical and Computer Engineering, University of Idaho, Moscow, ID 83844-1023, USA;
bguiana@uidaho.edu
* Correspondence: azadehgol@uidaho.edu; Tel.: +1-208-885-9000
† These authors contributed equally to this work.

Abstract: Electromagnetic (EM) scattering may be a significant source of degradation in signal and power integrity of high-contrast silicon-on-insulator (SOI) nano-scale interconnects, such as optoelectronic or optical interconnects operating at 100 s of THz where two-dimensional (2D) analytical models of dielectric slab waveguides are often used to approximate scattering loss. In this work, a formulation is presented to relate the scattering (propagation) loss to the *scattering parameters* (S-parameters) for the smooth waveguide; the results are correlated with results from the finite-difference time-domain (FDTD) method in 2D space. We propose a normalization factor to the previous 2D analytical formulation for the stochastic scattering loss based on physical parameters of waveguides exhibiting random surface roughness under the exponential autocorrelation function (ACF), and validate the results by comparing against numerical experiments via the 2D FDTD method, through simulation of hundreds of rough waveguides; additionally, results are compared to other 2D analytical and previous 3D experimental results. The FDTD environment is described and validated by comparing results of the smooth waveguide against analytical solutions for wave impedance, propagation constant, and S-parameters. Results show that the FDTD model is in agreement with the analytical solution for the smooth waveguide and is a reasonable approximation of the stochastic scattering loss for the rough waveguide.

Keywords: dielectric slab waveguide; discrete filtering; exponential autocorrelation; FDTD; optical interconnects; photonics; random roughness; stochastic scattering loss; scattering parameters; S-parameters

Citation: Guiana, B.; Zadehgol, A. Characterizing THz Scattering Loss in Nano-Scale SOI Waveguides Exhibiting Stochastic Surface Roughness with Exponential Autocorrelation. *Electronics* **2022**, *11*, 307. https://doi.org/10.3390/electronics11030307

Academic Editors: Giulio Antonini, Daniele Romano and Luigi Lombardi

Received: 24 November 2021
Accepted: 16 January 2022
Published: 19 January 2022

Publisher's Note: MDPI stays neutral with regard to jurisdictional claims in published maps and institutional affiliations.

1. Introduction

Nano-scale SOI optical interconnects, comprised of silicon/silicon-dioxide (Si/SiO$_2$) dielectric waveguides operating at 100s of terahertz (THz), constitute an increasingly important building block of modern integrated circuits, where the high-tech market demands smaller form-factors and wavelengths. Considering the non-ideal manufacturing process, random imperfections in the surfaces of nano-scale dielectric waveguides may cause significant signal degradation and power attenuation, as EM waves propagate through the interconnect structure, where the loss is primarily due to EM wave scattering with surface roughness of the waveguide [1–5]. Therefore, the characterization of scattering loss is a topic of significant interest to the scientific community [1–10].

The three-dimensional (3D) structure of SOI optical interconnects poses certain challenges to its analytical and numerical modeling; thus, the stochastic scattering loss observed in nano-scale THz SOI interconnects is often approximated using 2D planar models of the dielectric slab waveguide exhibiting surface roughness. The 2D analogue is useful for analytically [5,8,10] characterizing the effect of scattering loss on the power attenuation of light waves, and is used as a comparison for both experimental analysis (of physical waveguides) [1–4] and numerical analysis [1,2,6,7,9].

In 1983, Kuznetsov and Haus [11] published their work on using the 3D volume current method (VCM) to evaluate the radiation loss in dielectric waveguide structures. Their work includes analysis of single-line, two-line coupled, and three-line coupled waveguide structures, in the absence of random surface roughness. In 1990, Lacey and Payne [5] released their seminal work analyzing 2D planar waveguides exhibiting random surface roughness for a single-line waveguide structure. Their work applies Green's functions to the structure, operating in the transverse-electric-to-z (TE^z) mode, as an approximation for scattering loss in 3D optical interconnects, and later in 1994 [8] it was updated to use normalized waveguide parameters. In 2005, Barwicz and Haus [6] expanded on both of those developments by applying the 3D VCM to single-line waveguides exhibiting random surface roughness. In each of these cases, despite having relatively simple geometries, the solutions are formulated around complicated integral equations, and the solutions only become more complicated as the geometry becomes more complex, for example by adding roughness, multiple tightly-coupled lines, arbitrary-shaped lines or grating, etc., thus limiting the application of integral-based solutions. An effective workaround to the integral-equation complication is to reformulate the problem around differential-equations, leading to the FDTD method. A version of the FDTD method based on wavelets is used in [2], but the details of the FDTD formulation are not included. The FDTD method is also used in [1] through the software tool *Lumerical*, but again details of the FDTD methodology are absent.

The major contributions of the present work are as follows. (**1**) We provide the FDTD methodology for analysis of 2D dielectric waveguides exhibiting random surface roughness, operating in the TE^z mode. (**2**) We propose a methodology for the extraction of S-parameters, and we apply that methodology to the characterization of scattering loss. (**3**) We improve the computational efficiency of this model by using filtering techniques to attenuate numerical noise from simulation results, thereby allowing for the use of a relatively coarse spatial and temporal discretization while retaining the integrity of numerical results. (**4**) While the integral-based VCM has increasing complexity as the geometry becomes more complex, the FDTD method is especially well-suited for arbitrary waveguide geometries and arbitrary surface roughness profiles. To keep this presentation simple and concise, we chose to apply the methodology to a single line; however, it can easily be adapted to multiple tightly-coupled lines. (**5**) We provide the Python [12] code titled *Optical Interconnect Designer Tool* (OIDT) [13] which features multi-CPU-core support for parallelized FDTD, as an open-source software package [14] hosted on GitHub through a public repository under the GNU GPL v3.0 license [15], to encourage further exploration and (inter)national collaboration on optical interconnect research.

While the FDTD method implemented via a *serial* programming paradigm would be computationally expensive, its highly *parallelizable* nature may provide a potential path to a computationally expedient solution; thus, herein we begin to explore this potential by developing a parallelized implementation of FDTD with a traditional Yee-based algorithm [16,17] and convolution perfectly matched layer (CPML) [18] boundaries, to characterize the scattering loss in dielectric slab waveguides exhibiting surface roughness.

The remainder of this paper is organized as follows. Section 2.1 outlines the physical waveguide structure analyzed throughout this paper. Section 2.2 establishes the details of the FDTD model being used, where Section 2.2.1 provides additional details on the discretization and application of random roughness profiles to the FDTD environment. Section 2.2.2 details the filtering technique used to improve numerical measurements from FDTD simulations. Section 2.2.3 addresses the coordinate transformations used to move between the analytical solution and the FDTD model. In Section 2.3, we provide the methodology for S-parameter extraction in 2D FDTD. In Section 2.3.2, we formulate the relations between the S-parameter matrix and the scattering loss α. In Section 2.4, we propose an updated equation for computing the stochastic scattering loss α based on physically realistic waveguide parameters and define its components, including a discussion on the exponential ACF. In Section 2.5, we validate the FDTD model by correlating against analytical expressions for the wave impedance, the propagation constant, and the ideal S-parameter matrix. In Section 3, we discuss the numerical results of FDTD, and the potential sources of error between FDTD and the analytical model. In Section 3.1, we compare our results against those of other investigators. In Section 3.2, we examine the dependence of α on input power and formulate the mode normalization N_F. We conclude with closing remarks in Section 4.

2. Methods

2.1. The Waveguide Structure

Optical interconnects are comprised of spatially 3D waveguide structures with a certain surface roughness profile which may not vary much relative to the smooth (flat) waveguide's width. Often, 2D models of dielectric slab waveguides with the same height and similar material parameters are used to analyze the 3D waveguide. This allows the analysis to be decomposed into two modes: (1) *transverse electric* (TE) and (2) *transverse magnetic* (TM).

Here, we analyze the structure in Figure 1 for the TE mode. We start by defining the coordinate grid in the $\hat{x}$-$\hat{z}$ plane, and orient the device to operate with infinite extent in both the $\hat{y}$ and the $\hat{z}$ directions, where the waveguide length and power flow are along $\hat{z}$ and the waveguide height is infinite along $\hat{y}$.

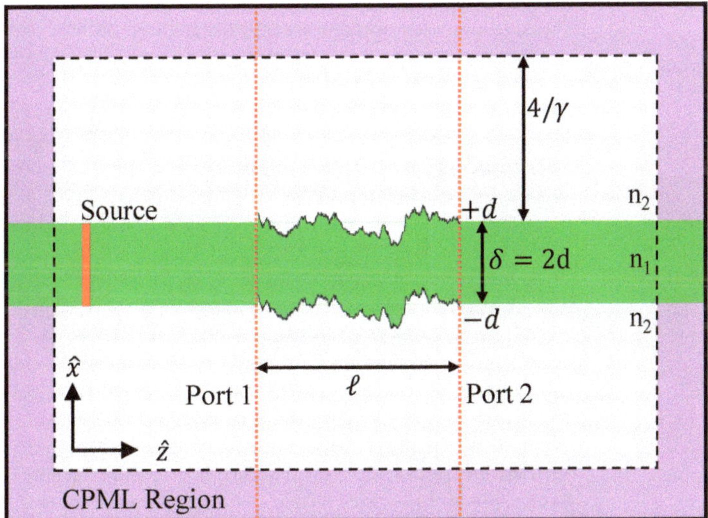

Figure 1. The baseline dielectric slab waveguide structure.

The dielectric slab waveguide consists of two regions, the *core* and the *cladding*. The core has a refractive index of n_1, and the cladding has a refractive index of n_2, where $n_1 > n_2$. The core region has a finite nominal width, which is typically denoted as two half-widths. Here, the width is $\delta = 2d$, where d is the half-width used throughout this work. The fields in the waveguide are assumed to be *time harmonic* (with $e^{j\omega t}$ dependence, where ω (rad/s) is the angular frequency) in nature with the E-field taking the form in (1).

$$\mathbf{E}(x,z) = \hat{y}\,\Phi(x)e^{-(\alpha+j\beta)z},\tag{1}$$

where $\beta = n_{\text{eff}}k_0$ (rad/m), n_{eff} is the effective index found via the effective index method (EIM) [10], k_0 is the free-space wave number, α is the attenuation constant resulting from sidewall roughness, and $\Phi(x)$ is a piece-wise function with its components defined in (2)–(4). In (2) the term A_e is a scaling constant.

$$\Phi(x) = \begin{cases} A_e \cos(\kappa x) & |x| \leq d \\ A_e \cos(\kappa d)e^{-\gamma(|x|-d)} & |x| > d \end{cases}, \tag{2}$$

$$\kappa^2 = n_1^2 k_0^2 - \beta^2 \tag{3}$$

$$\gamma^2 = \beta^2 - n_2^2 k_0^2 \tag{4}$$

Random perturbations exist along the boundary between the core and cladding, resulting in the *surface roughness profile*. We use an *exponential ACF* to describe the surface roughness by its standard deviation σ and its correlation length L_c. The *mean* of the profile is 0, so the random perturbations are not included in the nominal width. We further assume the cladding extends infinitely outward from the core region. The FDTD environment used in this paper is described fully in [7,14,19], and in Section 2.2.

2.2. The FDTD Environment

The nominal waveguide structure may expediently fit into 2D FDTD analysis. We start by converting the nominal structure into uniform discrete cells with side length Δx. The temporal resolution is then set at the *Courant stability limit* [20] for 2D FDTD, where the background material is set to the cladding medium. We apply the CPML [18] to the exterior of the computational domain, thereby simulating infinite space with minimal reflections and computational cost.

We define a length ℓ over which we generate and discretize a random profile; the remaining FDTD cells create an extra buffer space to allow for modal waves to settle, after leaving the source point and before reaching the recording (numerical measurement) point. Each source and recording location are designated by a separate *port*, e.g., a single optical line would be characterized by two ports, with one port at each end of the line.

Once the roughness profile is ready, it is applied as the core and cladding boundary between ports. Referring to Figure 1, we place a source condition along $\hat{x}$ in a vertical line of cells across the entire opening of the waveguide, where the distance between the source and CPML is more than 10 cells, while we may approximate infinite space with the CPML, we still need to retain a buffer space in the cladding between the waveguide and the CPML boundary. To capture the intricacies of the interactions of the EM fields in both the core and cladding regions, we need to set the cladding size appropriately. Therefore, it is necessary to capture as much of the E-field as possible. Note, in (2) the E-field magnitude decays exponentially in the cladding region with a rate of $1/\gamma$, and we can use that behavior to set the cladding buffer size. At a distance of $4/\gamma$ from the core/cladding boundary, the E-field magnitude at the edge of the simulation space is no more than 2% of the E-field magnitude at the core/cladding interface, and it only decays further from there; thus, the cladding region size is set accordingly, as in Figure 1.

Data are collected in the form of E-field values at ports 1 and 2 along the first line of cells adjacent to the rough region. These points are recorded at each time-step for the duration of the FDTD analysis. In post-processing, we take the recorded time-domain E-field values and convert them to the frequency domain with the fast Fourier transform (FFT) [21]. We then numerically integrate the E-field over the recorded line of cells, resulting in a frequency dependent voltage with which further analysis may be performed.

We set up the FDTD grid based on the waveguide geometry, material parameters, and desired frequency range. The geometry is set up as shown in Figure 1, where $n_1 = 3.5$ and $n_2 = 1.5$. We additionally set the fundamental frequency as $f_0 = 194.8$ THz (corresponding to source wavelength $\lambda_s = 1.54$ μm). Using the core refractive index, we find the minimum phase velocity $v_{\min}$ (m/s). Using the fundamental frequency, we assign our desired maximum frequency as $f_{\max} = N_H f_0$ (Hz), where N_H is the number of desired harmonics above the fundamental. Using both the minimum phase velocity and the maximum frequency, we find the minimum wavelength simulated in the FDTD scheme with $\lambda_{min} = v_{min}/f_{max}$. Then, our spatial discretization is $\Delta x = \lambda_{min}/30$. At $N_H = 2$, $\Delta x \approx 7.3$ nm/cell. We set the time-step Δt at the Courant limit based on the cladding material which has the largest possible phase velocity in the FDTD environment, such that $\Delta t = \frac{\Delta x}{\sqrt{2}v_{clad}}$.

The total grid is 5554 cells $\times$ 247 cells ($\hat{z} \times \hat{x}$) with 26,997 time-steps. We use 40 layers of CPML as the absorbing boundary condition because in our experience 30–40 CPML layers provides good correlation of wave impedance within 1%. Along $\hat{x}$, the core region is centered and the nominal full width measures 36 cells, where the remaining 300 cells are evenly distributed on either side of the core as cladding. Along $\hat{z}$, $\ell = 4092$ cells, and the remaining 1858 cells are evenly distributed to each port region.

The computations are done by using a workstation with two Intel® Xeon® E5-2687W v3 CPUs (40 logical cores), operating at 3.10 GHz. Each simulation occupies less than 410 MB of RAM and is completed in roughly 300 s (or 5 min).

2.2.1. Verifying the Validity of Discretized Roughness Profiles

We start by assigning a target for σ_o, L_{c_o}, and μ_o, where μ designates the *mean* in this subsection. These parameters are then normalized by the spatial discretization step-size Δx value used in the FDTD simulation to yield $\{\sigma = \sigma_o/\Delta x, L_c = L_{c_o}/\Delta x, \mu = 0\}$. The discrete values are passed into the *Pyspeckle* [22] Python library which uses the methods in [23] to generate random profiles; this generation process returns an array of a specified size with floating point values quantifying the surface perturbation. As was the case in [7], a linear offset is added to the aforementioned floating point array to ensure that all values are positive. The offset array is then cast to integer values via the *floor* function and the same linear offset is subtracted from the now integer array, where the final discrete array has parameters σ', L'_c, and μ'. The error between input (σ) and output (σ') parameters may be quite large, due to the discretization process. However, we may circumvent this issue by constraint-based generation of profiles, described below.

We set a percentage tolerance for the normalized input parameters $\{\sigma, L_c, \mu\}$ and we check that the output parameters $\{\sigma', L'_c, \mu'\}$ fit the input parameters within the prescribed tolerance. If a profile does not meet the criteria it is discarded and a new profile is generated. In our numerical experiments, the tolerance is specified by $\sigma' \in [0.9\sigma, 1.1\sigma]$ and $L'_c \in [0.9L_c, 1.1L_c]$, and $\mu' \in [-0.01, +0.01]$.

We find σ' and μ' via built-in Numpy functions *std* and *mean*, respectively. We may estimate the L'_c value that fits the autocorrelation data, as explained next. We start by finding the autocorrelation of the generated discretized surface profile using the Pyspeckle *autocorrelate* function which provides a normalized array with its maximum value occurring at $\zeta = 0$. Note the autocorrelation of the generated profile tracks an exponential ACF up to the correlation length, as can be seen in Figure 2. With that in mind, we apply a root finding technique to determine L'_c while using $R_{XX}(L'_c) = e^{-1}$ as the reference value. We then subtract e^{-1} from the discrete ACF and find the root closest to $\zeta = 0$, which is the correlation length of the discrete ACF. We may then compare the L'_c with L_c to determine the validity of the generated discretized profile.

For samples with parameter values $\Delta x = 11$ nm and $\sigma = 9$ nm, and knowing that the probability distribution function (PDF) of the random process is *normal* in nature, we know that 99% of values in the final array will be contained in the range $\pm 3\sigma \approx \pm 2.46$. Applying the *floor* function to this range results in the discrete set $\{-3, -2, -1, 0, 1, 2\}$ which may cause significant differences between output values $\{\sigma', L'_c, \mu'\}$ and input values $\{\sigma, L_c, \mu\}$.

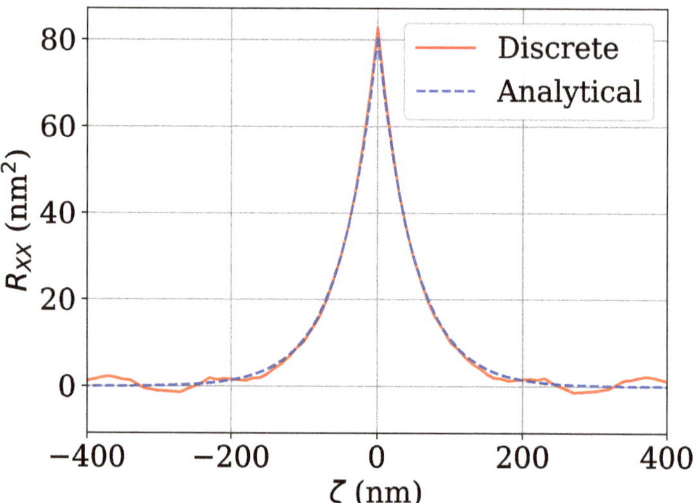

Figure 2. Example ACF with input parameters $\sigma = 9$ nm and $L_c = 50$ nm. The discrete trace is generated with 5000 samples.

Figure 2 compares a discretized exponential ACF comprised of 5000 samples against the continuous analytical (18). Even at this small sample size, the discretized profile still correlates well to the ideal ACF up to $\zeta = L_c$, but after that point, there is noticeable noise. At $\zeta = 0$ nm the discretized and continuous analytical ACFs do not line up perfectly. The misalignment at $\zeta = 0$ nm may be remedied by normalizing the analytical ACF to match the discretized ACF, using σ' instead of σ.

2.2.2. FDTD Noise Reduction

Numerical experiments involving waveguides with random surface roughness result in a certain amount of noise in α. We may observe that in the initial calculation, labeled unfiltered, there is rapid oscillation around the trend; such oscillations are undesirable and cause scattering loss readings to vary between numerical experiments. This issue may be resolved by the use of a *moving-average* function over the frequency range of interest; this technique is often used to reduce noise levels in digital signals [24]. We use (5) to reduce the noise level at each frequency point, where N is the number of samples to either side the reference index k.

$$\alpha_{\text{Filtered}}[k] = \frac{1}{2N+1} \sum_{p=-N}^{N} \alpha_{\text{Unfiltered}}[k+p] \tag{5}$$

The rolling average function is effective at reducing noise levels, but due to the smoothing effect it also introduces its own set of numerical distortions. However, the introduced error is small when N is small. As such, we use $N = 5$ to generate the filtered curve, where the rapid oscillations have been reduced but the trend remains mostly unchanged.

2.2.3. Modal Transformations and Coordinate Mapping

The geometry used for the characterization of scattering loss from random surface perturbations, shown in Figure 1, is based on the geometry from Figure 5 in [10]. The fields in [10] are described as TE^z, since $E_z = 0$ and $\frac{\partial}{\partial y} = \partial_y = 0$, and by using (56), (59), and (60) in [10] we know the non-zero field components are $\{E_y, H_x, H_z\}$, while $E_z = E_x = H_y = 0$. The TE^z field configuration may also be obtained from (6-72) in [25] and setting $\partial_y = 0$, and it is mathematically identical as two other modes; specifically, setting $\partial_y = 0$ in (6-64) in [25] yields the TM^y, and in (6-74) in [25] yields TE^x.

In our geometry of Figure 1, there exists a single nonzero E-field component along the invariant (infinite) direction ($\hat{y}$), and two nonzero H-field components along the finite directions ($\hat{z}$, $\hat{x}$) which may be interpreted as either height or width [10]. Out of the three mathematically equivalent modes (TE^z, TM^y, TE^x), we choose the TE^z field configuration here, as it aligns best with the physical interpretation of the physical waveguide with propagation along $\hat{z}$ (length), a transverse E-field along $\hat{y}$ (width or height), and H-field components along $\hat{x}$ and $\hat{z}$.

Our FDTD simulations are based on the traditional Yee algorithm in a 2D lattice, as formulated in ([17], ch. 3). Our FDTD formulation is derived with the assumption that $H_z = 0$ and $\partial_z = 0$, resulting in the TM^z mode with field components $\{E_z, H_x, H_y\}$. This FDTD lattice may initially appear to be in conflict with our analytic formulation; however, note that the E-field has a single nonzero component along the infinite (invariant) direction, and the H-field has two nonzero components. Since we can assign the FDTD geometry in an arbitrary manner, we choose to orient H_x along the length and H_y along the width (or height), resulting in a field configuration with the same orientation as the analytical formulation but with a rotated coordinate grid. We can rotate the coordinate grid of the analytical field configuration such that it results in a configuration *identical* to the FDTD fields by the steps shown in Figure 3.

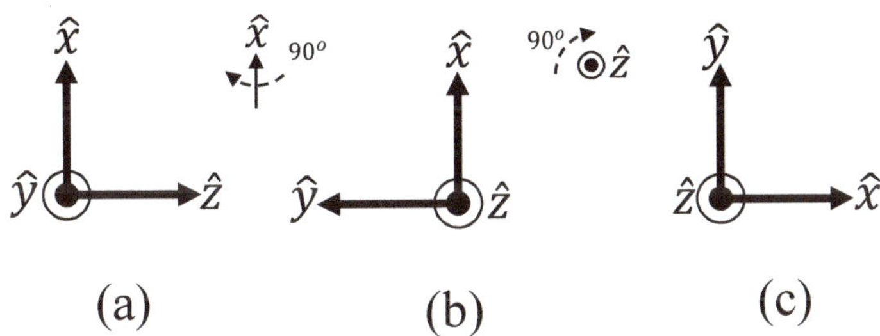

Figure 3. Coordinate grid rotation steps. (**a**) Initial orientation of analytical formulation for 2D TE^z. (**b**) Intermediary rotated mapping. (**c**) Final mapped orientation of analytical formulation is identical to the FDTD formulation for 2D TM^z.

In Figure 3, starting with the analytical expression in (a), we rotate the coordinate grid twice. The first rotation is 90° around $\hat{x}$ from $\hat{z}$ toward $\hat{y}$; this produces the grid in (b). The second rotation is 90° around $\hat{z}$ from $\hat{y}$ to $\hat{x}$; this produces the grid in (c). The mapping is complete after these rotations, and we can then use the FDTD 2D TM^z field components $\{E_z, H_x, H_y\}$ to represent the analytical 2D TE^z field components $\{E_y, H_z, H_x\}$, respectively, with no modifications to the established FDTD formulation nor the analytical formulation.

2.3. S-Parameters Extraction Methodology in FDTD

S-parameters are often used to characterize a variety of electronic systems [7,19,26]. The methodology of finding S-parameters may be applied to 2D FDTD simulations quite expediently [7,19].

We use the traditional definition of S-parameters [26], where the *total* voltage wave measured at each port in a system can be decomposed into *incident* and *reflected* waves, i.e., $\tilde{V} = \tilde{V}^+ + \tilde{V}^-$, and those components can be used to evaluate S-parameters as in (6), where m, n, k are port numbers, $\tilde{V}^+$ is the incident wave, $\tilde{V}^-$ is the reflected wave, and $\tilde{V}$ is the total wave.

$$S_{mn} = \left.\frac{\tilde{V}_m^-}{\tilde{V}_n^+}\right|_{\tilde{V}_k^+ = 0 \ \forall k \neq n} \tag{6}$$

In our FDTD simulations, we use a two-port system, so $\{m, n, k\} \in \{1, 2\}$. This methodology may be further extended to systems with more than two ports.

2.3.1. Computing S-Parameters, Using FDTD

We are able to record total, incident, and reflected fields in FDTD simulations, but these may not all be recorded simultaneously. Therefore, we utilize a *four-step simulation setup* for collecting S-parameters.

Other simulations for calculation of loss may be considered a subset of the four-step simulation process [19]. Each of the simulation steps are visualized in Figure 4, and explained below. The *baseline* setup and geometry is in Figure 1.

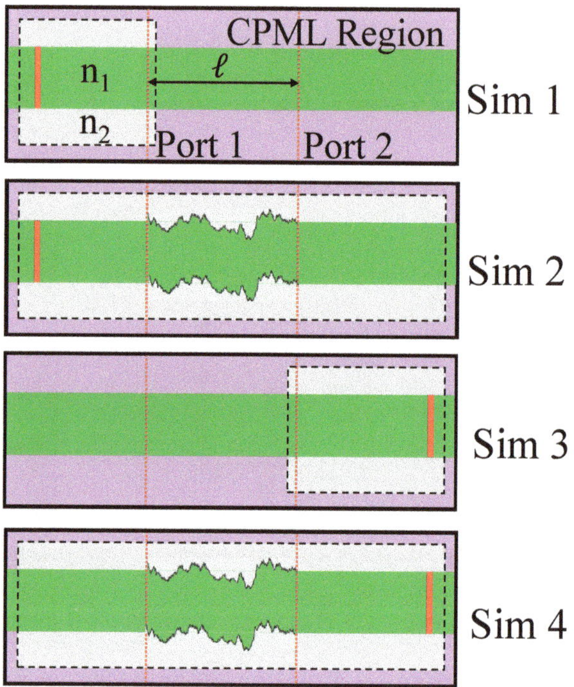

Figure 4. Four-step simulation setup for S-parameters extraction methodology.

Sim 1: The first simulation starts by placing the source condition at port 1, so $\tilde{V}_2^+ = 0$ for this simulation. We simplify the geometry of the dielectric slab by excluding the random sidewall perturbations at this time. Additionally, we extend the right-side CPML boundary to 10 cells to the right of port 1. This simulation results in $\tilde{V}_1^+$.

Sim 2: The second simulation starts with the same source condition as simulation 1. We then apply a valid discrete roughness profile to top and bottom boundaries between core and cladding. In our simulations, we chose the top and bottom profiles to be identical, but other choices are possible too. The CPML boundaries are evenly distributed around the computational domain. This simulation results in $\tilde{V}_1$ and $\tilde{V}_2^-$.

Sim 3: The third simulation is similar to the first simulation. We place the source condition at port 2, where $\tilde{V}_1^+ = 0$ in this simulation, and extend the left-side CPML boundary to 10 cells to the left of port 2. Simulation 3 results in $\tilde{V}_2^+$.

Sim 4: The fourth simulation finalizes the port field data collection. Similar to the third simulation, we place the source condition at port 2, and similar to the second simulation we set the CPML boundaries at the baseline limits and apply the roughness profile in the same manner. This simulation results in $\tilde{V}_1^-$ and $\tilde{V}_2$.

The fields recorded in the numerical experiments are limited to the four aforementioned steps, but there are still two field components missing which would fully describe the scattering matrix; those are $\tilde{V}_1^-$ when $\tilde{V}_2^+ = 0$, and $\tilde{V}_2^-$ when $\tilde{V}_1^+ = 0$. Here, we may use the decomposition relation to find the implicit reflected fields. Specifically, using $\tilde{V}_1^+$ from Simulation 1 and $\tilde{V}_1$ from Simulation 2, we may obtain $\tilde{V}_1^- = \tilde{V}_1 - \tilde{V}_1^+$. Similarly, we may recover $\tilde{V}_2^-$ from simulations 3 and 4.

With each of these values calculated from the numerical experiments, the S-parameters matrix may now be computed. In ideal waveguide with no sidewall perturbations, we would expect the S-parameters matrix (of the waveguide system) to be both *symmetric* (*reciprocal*) and *unitary* (lossless) [26]. For a waveguide with sidewall perturbations, we expect the S-parameters matrix (of the waveguide system) to be symmetric (reciprocal) but non-unitary (lossy) [26]. Higher port numbers may be simulated by following the same process of incident field simulation followed by total field simulation, for each port subsequently.

2.3.2. Direct Method of Computing Scattering Loss, Using FDTD

Using the second simulation, we may find the *total* voltage wave values at both ports 1 and 2, but this time we label them $\tilde{V}(0)$ and $\tilde{V}(\ell)$, respectively. We assume the voltages to have the same form as the electric field but with the $\Phi(x)$ element replaced with V_0, i.e., $\tilde{V}(z) = V_0 e^{-(\alpha + j\beta)z}$. In this form, it may be observed that as z increases, we also expect the voltage to attenuate in amplitude and accumulate in phase.

Since we have measured the voltage at two port locations, we may determine the attenuation. To do this, we divide $\tilde{V}(\ell)$ by $\tilde{V}(0)$, resulting in (7).

$$\frac{\tilde{V}(\ell)}{\tilde{V}(0)} = \frac{V_0 e^{-(\alpha + j\beta)\ell}}{V_0} = e^{-(\alpha + j\beta)\ell} \tag{7}$$

Using (7) we may isolate α by using the complex logarithm where $z \in \mathbb{C}$, $\log(z)$ is the complex-domain natural logarithm of z, $\ln(|z|)$ is the natural logarithm with base e, and $\arg()$ is the true angle of z; i.e., the angle of z which includes all full turns and may have a magnitude greater than $\pm\pi$ [27].

$$\log(z) = \ln(|z|) + j\arg(z), \tag{8}$$

Applying (8) to (7) results in (9).

$$-\alpha\ell - j\beta\ell = \ln\left(\left|\frac{\tilde{V}(\ell)}{\tilde{V}(0)}\right|\right) + j\arg\left(\frac{\tilde{V}(\ell)}{\tilde{V}(0)}\right) \tag{9}$$

Equation (9) may be separated into real and imaginary components, resulting in the final expression in (10), for calculating power loss directly from FDTD experiments.

$$\alpha = -\frac{1}{\ell}\ln\left(\left|\frac{\tilde{V}(\ell)}{\tilde{V}(0)}\right|\right) \quad (\mathrm{Np/m}) \tag{10}$$

While (10) is possibly the most direct method for calculating power loss from FDTD simulation, there is an alternative definition which could accomplish that task through the use of S-parameters. We start by taking the argument of the natural logarithm in (10) and squaring it, but instead of 0 and ℓ being the reference points, the voltages are now in reference to ports 1 and 2, leading to

$$A = \left|\frac{\tilde{V}(\ell)}{\tilde{V}(0)}\right|^2 = \left|\frac{\tilde{V}_2}{\tilde{V}_1}\right|^2 = \left|\frac{\tilde{V}_2^+ + \tilde{V}_2^-}{\tilde{V}_1^+ + \tilde{V}_1^-}\right|^2. \tag{11}$$

In (11), we may replace the magnitude-squared operation with the equivalent complex operation, resulting in

$$A = \frac{(\tilde{V}_2^+ + \tilde{V}_2^-)(\tilde{V}_2^+ + \tilde{V}_2^-)^*}{(\tilde{V}_1^+ + \tilde{V}_1^-)(\tilde{V}_1^+ + \tilde{V}_1^-)^*}, \tag{12}$$

where * denotes the complex conjugate operator.

Simplifying (12), we may combine the incident and reflected voltage waves into compact S-parameters from. We may then reinsert A into a natural logarithm and recover the expression for α as a function of S-parameters in (13) [19].

$$\alpha = -\frac{1}{\ell}\ln\left(\left|\frac{(1+S_{11})(S_{12})}{S_{11}+S_{12}S_{21}}\right|\right) \quad \text{(Np/m)}. \tag{13}$$

The form of α in (13) may also be utilized on systems with larger number of ports. In this case, port 1 is used as the reference port for loss, but in general the reference port may be any port in a multi-port system. In a large multi-port system, the loss equation using S-parameters may be computed for each reference port.

2.4. Analytical Loss Function

In the ensuing formulation, we assume simple media; i.e., linear, isotropic, and non-dispersive. Following the recent work in [10], we propose (14) as the generic scattering loss function.

$$\alpha = \frac{1}{N_F}\Phi^2(d)M_W S_W \quad \text{(Np/m)}. \tag{14}$$

Equation (14) is similar to the loss function used in previous works [5,8,10], but we have made a modification by dividing α in those works by the normalization factor N_F.

Either part of the piece-wise function (2) may be evaluated at $x = d$ and result in (15).

$$\Phi^2(d) = A_e^2 \cos^2(\kappa d), \tag{15}$$

Note that A_e, derived in Equation (73) in [10], is dependent on the input power. This dependency is not desirable; thus, it is eliminated by introducing the factor, N_F, in (14), normalizing the amplitude of α, as explained in Section 3.2.

The term M_W is defined with (16).

$$M_W = \left(n_1^2 - n_2^2\right)^2 \frac{k_0^3}{4\pi n_1}. \tag{16}$$

The term S_W is defined with (17) and represents the contribution by the surface roughness described by a random distribution.

$$S_W = \int_0^\pi \tilde{R}_{XX}(\beta - n_2 k_0 \cos(\theta))d\theta, \tag{17}$$

where $\tilde{R}_{XX}$ is the power spectral density of the surface roughness profile.

Surface roughness may be approximated as a stationary random process [10], therefore the power spectral density may be recovered through the ACF of the roughness profile, which may be assumed to have an *exponential* shape [4,5]. The exponential ACF is given by (18) [10]

$$R_{XX}(\zeta) = \sigma^2 e^{-\left|\frac{\zeta}{L_c}\right|}, \tag{18}$$

where σ and L_c are the standard deviation and correlation length of the profile, respectively, and ζ is the spatial shift variable.

There are two observations that could be made about (18) by setting ζ to specific values. First, $R_{XX}(\zeta = 0)$ results in the variance of the profile. Second, $R_{XX}(\zeta = L_c) = \sigma^2 e^{-1}$; this second result is used later for approximating the value of L_c from surface profile data generated by the *Pyspeckle* software [22]. Some photo-lithographic processes for Si/SiO$_2$ may lead to profiles with $\sigma = 9$ nm and $L_c = 50$ nm [4].

We define the spatial Fourier transform (SFT) as

$$\tilde{R}_{XX}(k) = \int_{-\infty}^{\infty} R_{XX}(\zeta)e^{-jk\zeta}d\zeta, \tag{19}$$

where the input function is translated from ζ-space (m) to k-space (rad/m), k is the wave number, and the imaginary number $\jmath = \sqrt{-1}$.

Applying the SFT (19) to (18), yields

$$\tilde{R}_{XX}(k) = \frac{2L_c\sigma^2}{1 + L_c^2 k^2}.$$ (20)

We may insert (20) into (17) and numerically evaluate the integral to obtain the contribution of the surface roughness profile to the loss α.

2.5. FDTD Model Validation

We validate the FDTD model being used in these numerical experiments in three ways, described below. Unless stated otherwise, the data in this section is generated for a smooth waveguide with $\delta = 300$ nm, $n_1 = 3.5$, $n_2 = 1.5$, and $f_0 = 194.8$ THz. Validation must be done prior to performing numerical experiments, so we utilize the smooth waveguide and the below methods for model validation. Using known and expected attributes of the smooth waveguide, we can compare the results obtained from numerical experiments to provide confidence in the validity of our model before performing numerical experiments with waveguides that exhibit surface roughness.

2.5.1. Wave Impedance

One such method is through comparison with a known analytical solution to the smooth dielectric slab waveguide. We use the wave impedance of an outward traveling wave. This solution is well established and has been derived in several places [10,25]. We find the wave impedance by dividing TE^z mode E-field by the corresponding H-field component along the length of the slab waveguide. Here, those fields are E_y and H_z, respectively. In the smooth slab case, the real portion of the wave impedance should be very small. For the analytical solution, the division between the E-field and H-field gives

$$Z_w = \jmath \frac{x}{|x|} \frac{\omega\mu}{\gamma} \quad (\Omega),$$ (21)

where μ is the magnetic permeability. Division of x by its magnitude is used to set the appropriates sign for either above or below the slab.

The FDTD portion of this comparison may be conducted through the second simulation from Section 2.3.1 with the surface roughness omitted. Since the wave impedance is calculated for an outward traveling wave, we use the E-field and H-field data in the cladding region. We take all the steps necessary to compare frequency-domain voltages as described in Section 2.2, but we exclude the final integration such that we are left with field data for every point along the line at ports 1 and 2. Using the port 2 data allows for the wave to propagate over a long enough distance to be well-set into the lowest order mode. We take the measurements from two cells below the lowest core cell which leaves a single cell buffer between the core region and the cell used for this calculation. Finally, the imaginary component is compared to the analytical solution.

The wave impedance calculated from the FDTD model is shown in Figure 5a. We can see that the impedance found from numerical experiment matches with the expected analytical value throughout this range of frequency samples. At f_0 the difference between the FDTD and analytical values is approximately 1.5 Ω, which translates to an error of less than 2% near the frequency of interest.

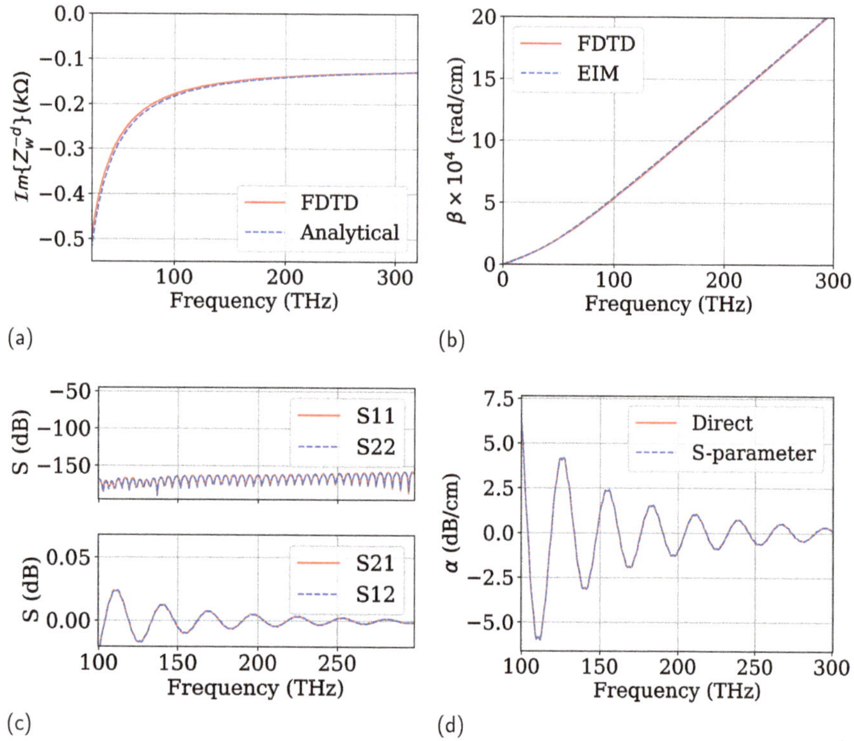

Figure 5. (**a**) Zeroth order TE^z mode wave impedance for the smooth dielectric slab waveguide. (**b**) Propagation constant β vs. frequency. (**c**) S-parameters for the smooth waveguide. (**d**) Propagation loss α (dB/cm) vs. frequency, for a smooth waveguide with S-parameters method vs. direct method for calculating propagation loss.

2.5.2. Propagation Constant

In the next validation method, we compare the propagation constant β obtained from FDTD against that obtained from the EIM in the frequency range of interest, at the same samples as the FDTD model output. We find β from FDTD by evaluating the imaginary component of (9).

Since we are examining the phase angle of the voltage measurements, the division of $\tilde{V}(\ell)$ by $\tilde{V}(0)$ may be converted to a subtraction, resulting in (22).

$$\beta = -\frac{1}{\ell}\left(\arg\left(\tilde{V}(\ell)\right) - \arg\left(\tilde{V}(0)\right)\right) \tag{22}$$

In FDTD, we compute the angle of the complex voltages over the entire frequency range on the smooth dielectric slab waveguide and unwrap the final angle array into full angle form (rather than principal angle form).

In Figure 5b, the traces are nearly overlapped. The fundamental frequency is highlighted by the vertical dashed line, where the error between the EIM and the FDTD model is less than 1%. This data further validates the FDTD model, and confirms the formulation leading to (22).

2.5.3. Scattering Matrix

In the last validation method, we utilize the properties of the S-parameters matrix described in Section 2.3.1. As stated, the matrix should be symmetric and unitary for the smooth waveguide. We extract the S-parameters from the FDTD model using the method presented in Section 2.3, and show these in Figure 5c.

Two observations are noteworthy in the frequency range of interest: (1) the cross-terms (S12, S21) have a magnitude of near 0 dB, indicating that there is almost complete transmission of power from one end of the waveguide to the other, and the self-terms (S11, S22) are correspondingly very small compared to the cross-terms, with a peak value of less than -150 dB; therefore, the matrix is nearly unitary as expected for a smooth lossless waveguide. (2) the S-parameter matrix is symmetrical, given the near perfect overlap of S11 with S22, and S12 with S21. These observations are noteworthy because they are the expected results for an ideal network, such as a smooth dielectric slab waveguide. Since the FDTD results align well with the expected behavior of a 2-port network, these results further validate the FDTD model.

We compare the *S-parameters method* of Section 2.3.1 and the *direct method* of Section 2.3.2 for calculating loss, as shown in Figure 5d. Here, we observe an oscillatory behavior similar to that in the cross-terms of Figure 5c. The oscillations hover around $\alpha = 0$ dB/cm and decay with increasing frequency, while the expected per-unit-length attenuation for an ideal smooth waveguide is $\alpha = 0$ dB/cm. Note that the loss from the S-parameters and from the direct method match very well, where the mean-squared error is on the order of 10^{-8}.

3. Results and Discussion

Unless stated otherwise, the data in this section are generated for waveguides with $\delta = 200$ nm, $n_1 = 3.5$, $n_2 = 1.5$, $f_0 = 194.8$ THz, and $\Delta x = 7.33$ nm.

In Figure 6, we show an example loss curve simulated in FDTD, to illustrate the need for filtering the FDTD output. As we can be seen in the figure, there is a nontrivial level of noise on the full range of α.

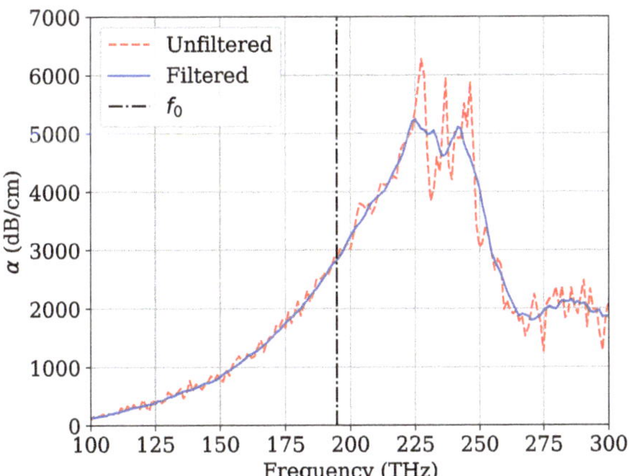

Figure 6. α vs. frequency, for a rough waveguide ($\sigma = 15$ nm, $L_c = 200$ nm) with noisy FDTD data (red dashed line) compared to filtered FDTD data (blue solid line), where f_0 is the excitation source frequency.

Applying the filter described in Section 2.2.2 results in the *Filtered* trace which considerably reduces the noise in the FDTD data. This is best exemplified by the α values for frequencies above 225 THz, where the noise is reduced by an order of magnitude. The α values from FDTD are subjected to this filtering technique prior to calculation of the percentage error between the FDTD and the analytical solution (14).

Figure 7 uses the data from tables I and II in [7], respectively, to illustrate the distribution of percent error between FDTD and analytical calculations. Figure 7a,b were generated with the standard deviation $\sigma = 9$ nm. The correlation length L_c varies uniformly in the range 200 nm to 1000 nm. The figures show the distribution of percentage errors for all correlation lengths with the same standard deviation, where a total of 924 roughness profiles were simulated using the FDTD model. We use these data to illustrate the effect of filtering on simulation results. In Figure 7a the mean error is −5.12%, whereas in Figure 7b the error is reduced to −4.12%. Likewise, the standard deviation reduces from 21.96 to 19.87%.

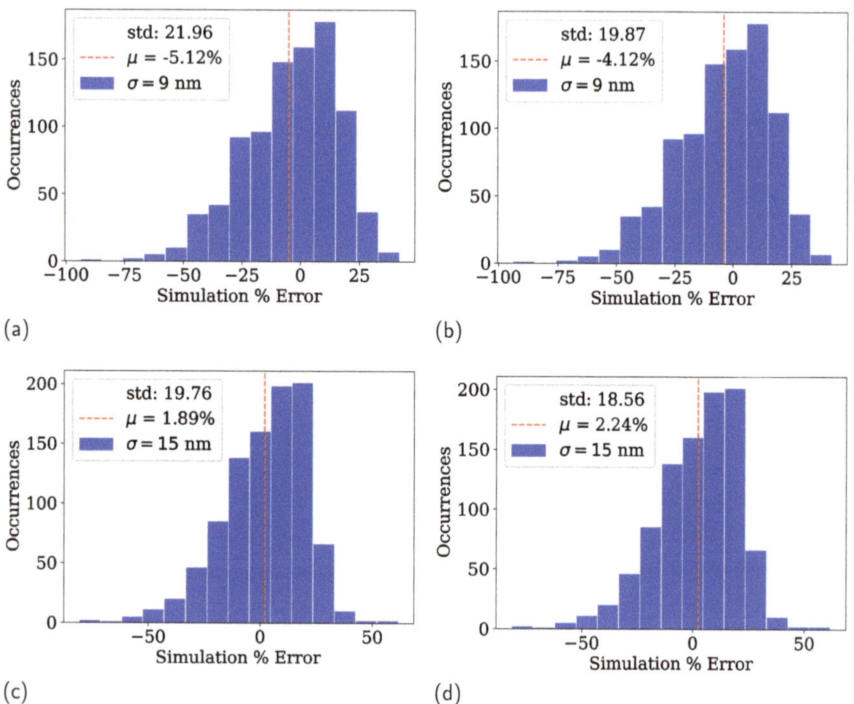

Figure 7. Percent error in propagation loss α (dB/cm) between analytical (14) vs. FDTD solutions: (**a**) 924 roughness profiles at $\sigma = 9$ nm, (**b**) same as (**a**) with data filtering, (**c**) 947 roughness profiles at $\sigma = 15$ nm, and (**d**) same as (**c**) with data filtering.

A similar setup is used for Figure 7c,d, where $\sigma = 15$ nm and L_c is uniform in the range 200 nm to 1000 nm, with a total of 90 roughness profiles simulated with the FDTD model. In Figure 7c the mean error is 1.89%, whereas in Figure 7d the error is increased, to 2.24%. Like in the $\sigma = 9$ plots, the standard deviation reduces, this time from 19.76 to 18.56%. From the numerical experiments conducted in FDTD on the relatively short-length slab waveguide, we have created the histogram of occurrences, which may be easily translated to a *probability mass distribution*.

Some potential sources of error in Figure 7 are listed below. (1) The parameters $\{\sigma', L_c', \mu'\}$, rather than $\{\sigma, L_c, \mu\}$, are used in the analytical solution when calculating the percentage errors. (2) We use a relatively coarse spatial and temporal resolution in the FDTD model, while utilizing a finer resolution grid may decrease the percentage error range, it would increase computation time. (3) We use (14) as the analytical model, which is based on the formulation originally proposed in [5] that used various approximations and simplifications, such as a first-order Taylor series expansion to evaluate the E-field.

From each setup in Figure 7 the maximum difference between filtered and unfiltered data is standard deviation from Figure 7a,b which reduces by 2.09%, and each other relevant value changes by around 1%, with the mean percentage error in the $\sigma = 15$ nm case actually increasing by 0.35%. This shows that the filtering step does not significantly impact the percentage error of the FDTD results from the analytical solution.

We define the *improvement metric* F_{imp} as

$$F_{\text{imp}}\left(x_{\text{Filtered}}, x_{\text{Unfiltered}}\right) = \left(1 - \frac{|x_{\text{Filtered}}|}{|x_{\text{Unfiltered}}|}\right) \times 100\%, \tag{23}$$

where $x_{\text{Filtered}} = \alpha_{\text{Filtered}} - \alpha_{\text{Analytical}}$, $x_{\text{Unfiltered}} = \alpha_{\text{Unfiltered}} - \alpha_{\text{Analytical}}$, α_{Filtered} and $\alpha_{\text{Unfiltered}}$ are defined in Section 2.2.2, and $\alpha_{\text{Analytical}}$ is defined by (14).

We use (23) to compare the effect of filtering on the FDTD data. Using this metric, we consider any value greater than 0% to be an improvement, and 100% would indicate the filter has maximally improved the unfiltered signal. We consider $\pm 5\%$ to be negligible change, i.e., filtering had little effect on unfiltered signal. For $\sigma = 9$ nm, filtering improves the error in 487 instances out of 924, and that filtering has negligible effect for 62 instances. For $\sigma = 15$ nm (23) results in improvement for 462 instances out of 947, and negligible effect for 107 instances.

It is worth noting that the above technique involves filtering α *after* E- and H-fields have been computed in FDTD. The data show that filtering in this manner does not add significant value. Thus, we plan to investigate filtering techniques within FDTD *during* the computation of E- and H-fields, as was demonstrated in [28] for suppressing spurious noise waves due to sub-gridding.

3.1. Comparing Our Results to Previous Publications

We compare this work to previous theoretical and experimental works. Equation (14) is based on the formulation provided in [10], which in turn is based on both [5,8]. The primary difference between [10] and this work is that here we introduce the normalization factor N_F. In [10], the components for α are based on the more physically realistic form of E-field (2) where A_e may be any real-valued scalar. The absence of the normalization factor (i.e., setting $N_F = 1$) leaves a dependency on input power; however, the input power may be modified to fit the expected scattering loss value for any point, by setting (24) equal to (25) and solving for α, which will always result in (29). In [5], the original formulation for finding scattering loss was proposed, where the mathematical normalization of (24) is used.

The follow-up work of [8] proposed a formulation for scattering loss calculations by using normalized waveguide parameters, but as was noted in [6], there appears to be an extra factor of 2 in the formulation used therein. We can see this numerically by comparing [8] with this work, also in Table 1, where the loss value is double what is calculated by (14). We further see that an input power of $P_{\text{TE}} = 4.3$ mW in the formulation found in [10] is most similar to the normalized values found here, but other choices of input power (e.g., $P_{\text{TE}} = 1$ mW or $P_{\text{TE}} = 1.45$ mW) could yield different values for loss as they may not eliminate dependence of α on input power.

The VCM was used in [6] to verify the scattering loss calculations. Their work was done on several 3D waveguides, and these provide a similar analogue to the 2D structure simulated in this work.

Looking at the experimental side, in [3] a physical 3D dielectric optical interconnect was tested for scattering loss. Their results show that the 2D planar model [8] is generally an *overestimate* of what can be expected from physical hardware, and the 3D simulations in [6] are generally an *underestimate*. Furthermore, in [3], unit variance is used, making it unique compared to other experimental data and included in Table 1. Other experiments conducted on physical hardware include those of [1,4], where a scattering loss magnitude of ≈ 35 dB/cm is reported. These experiments were conducted on 3D SOI optical interconnects consisting of Si core and SiO_2 cladding extending 1 μm in each direction around the core, making them amenable to comparison with the 2D planar approximation. The loss value in [1,4] is approximately 36% of the loss value in (14), and approximately 72% of the loss value in [6].

Table 1. α values in (dB/cm).

Source (a: hardware experimental) (b: numerical/analytical)	$d = 210$ nm $L_c = 20$ nm $\sigma^2 = 1$ nm^2	$d = 250$ nm $L_c = 50$ nm $\sigma^2 = 81$ nm^2
a [4] Lee	N/A	≈ 34
a [1] Jaberansary	N/A	≈ 33
a [3] Horikawa	≈ 0.5	N/A
b [6] Barwicz	N/A	48.6
b [8] PL 94	1.87	193.7
b [10], $P_{TE} = 1$ mW	0.22	22.1
b [10], $P_{TE} = 1.45$ mW	0.32	32.1
b [10], $P_{TE} = 4.3$ mW	0.94	95.2
This work (14)	0.94	96.8

3.2. Mode Normalization

Although allowing α to have dependency on waveguide's structural attributes such as the surface roughness profile, material parameters, or even input wavelength is desirable, dependence on input power is not. In previous works [5,8], the E-field is mathematically normalized, such that

$$\int_{-\infty}^{\infty} \Phi^2(x)\, dx = 1. \tag{24}$$

While the above normalization removes the dependence of α on the input power, it also forces $\Phi(x)$ to satisfy a certain *mathematical* requirement; however, if we evaluate (24) by inserting the *physical* field expression (2) for the TE modes, we arrive at (25) [10].

$$\int_{-\infty}^{\infty} \Phi^2(x)\, dx = A_e^2 \left(d + \frac{1}{\gamma} \right). \tag{25}$$

Note that if (24) is assumed to be true, then the field amplitude A_e would be forced to have the particular value $A_e^2 = 1/(d + \frac{1}{\gamma})$; however, this assumption may not hold true in the physical waveguide. To remedy the inconsistency, we start with [5]

$$\alpha = \frac{P_{rad}/2L}{P_g} = \frac{\frac{n_2}{2\eta_0} \int_0^{\pi} \frac{\langle |E_x(r,\theta)|^2 \rangle}{2L} r\, d\theta}{\frac{n_1}{2\eta_0} \int_{-\infty}^{+\infty} \Phi^2(x)\, dx}, \tag{26}$$

where the core and cladding refractive indices are tied to the guided power P_g and the radiated power P_{rad}, $\langle f \rangle$ designates the ensemble average of f [10], and $\eta_0 \approx 377\ \Omega$ [5].

We use (15) from [5] to simplify the numerator in (26) and note that $2\eta_0$ cancels at this point, resulting in (27).

$$\alpha = \frac{n_2}{n_1} \frac{\Phi^2(d)\left(n_2^2 - n_1^2\right)^2 \frac{k_0^3}{4\pi n_2} \int_0^{\pi} \tilde{R}_{XX}(\beta - n_2 k_0 \cos\theta)\, d\theta}{\int_{-\infty}^{+\infty} \Phi^2(x)\, dx}. \tag{27}$$

We use the definition of M_W and S_W here to further simplify (27); this results in (28), where n_2 in the numerator cancels and the n_1 in the denominator is pulled into M_W.

$$\alpha = \frac{\Phi^2(d) M_W S_W}{\int_{-\infty}^{+\infty} \Phi^2(x)\, dx} \tag{28}$$

We may further simplify (28) for the zeroth-order mode by inserting the result from (25) and the physical field amplitude expression of $\Phi^2(d)$ in (15), followed by canceling the resulting A_e^2 term in the numerator and denominator.

$$\alpha = \frac{\cos^2(\kappa d) M_W S_W}{d + 1/\gamma} \tag{29}$$

Therefore, to remove dependence on input power uniformly for any initial input field amplitude, we propose to define N_F in (14) as (30), where N_F has the same dependence on input power as $\Phi^2(d)$.

$$N_F = A_e^2 \left(d + \frac{1}{\gamma} \right), \tag{30}$$

Note that here, (29) is the simplest form for even TE modes, using the physical considerations based on actual field values in the waveguide. In [5], (24), is assumed to be true prior to calculating α, forcing the denominator of (28) to be 1.0 and limiting $\Phi^2(d)$ to be single valued. The crucial difference in the proposed formulation here (based on inclusion of N_F in α) lies in the assumption that the E-field amplitude can be functionally *any* value. We believe that the proposed normalization based on N_F offers a more physically consistent expression of α, and works more intuitively for correlation of numerical (or physical) experiments in FDTD (or in the lab) where the initial field amplitudes may be specified with regard to considerations independent of scattering loss.

4. Conclusions

Based on physical waveguide parameters, an explicit normalization factor (30) for the scattering loss α (14) was proposed. The equation was then used to compare the results across several previous publications, including both numerical and physical experiments, showing that the analytical equation is generally an overestimation of actual propagation loss in a physical waveguide. We used the proposed analytical equation to confirm the presence of an extra factor of 2 in [8].

The proposed analytical formulation of scattering loss was verified, using an expedient FDTD scheme which included extraction of the attenuation coefficient and S-parameters. We validated the FDTD scheme by comparing numerical results against previously published analytical functions for the dielectric slab waveguide.

With the FDTD model verified, we demonstrated S-parameters extraction and attenuation coefficient calculation by applying the proposed methodology to a smooth dielectric slab waveguide. We then applied the methodology to compute the attenuation coefficient for a dielectric slab waveguide exhibiting random sidewall perturbations according to the exponential autocorrelation function. We proposed to use a filtering technique to reduce the signal-to-noise ratio of the final FDTD data in the frequency domain.

Along the way, we demonstrated the ability of the FDTD scheme to produce reasonably accurate results through tens of simulations for sidewall roughness profiles of varying correlation length at standard deviations of $\sigma = 9$ nm and $\sigma = 15$ nm. The FDTD results showed that the mean error for simulation is quite small, with an overall average error of only -4.12% and 2.24% for the attempted standard deviations, respectively.

The Python FDTD code (OIDT) [13] used to generate much of the data in this paper was released as open-source software (under the GNU GPL v3.0 license [15]) published on GitHub [14], featuring multi-core support (for CPU) to compute the 2D TE^z fields of optical interconnects. Work is currently underway to develop efficient 2D transverse *magnetic* (TM), and full 3D, FDTD models for further characterization of stochastic scattering loss due to sidewall roughness in nano-scale optical interconnects, consisting of single-line and multiple tightly-coupled lines, operating at 100 s of THz.

Author Contributions: Conceptualization, A.Z.; Data curation, B.G.; Formal analysis, B.G. and A.Z.; Funding acquisition, A.Z.; Investigation, B.G. and A.Z.; Methodology, B.G. and A.Z.; Project administration, A.Z.; Resources, A.Z.; Software, B.G.; Supervision, A.Z.; Validation, B.G.; Visualization, B.G.; Writing—original draft, B.G.; Writing—review & editing, A.Z. All authors have read and agreed to the published version of the manuscript.

Funding: This work was funded, in part, by the NSF; Award # 1816542 [13].

Institutional Review Board Statement: Not applicable.

Informed Consent Statement: Not applicable.

Data Availability Statement: Not applicable.

Conflicts of Interest: The authors declare no conflict of interest.

References

1. Jaberansary, E.; Masaud, T.M.B.; Milosevic, M.M.; Nedeljkovic, M.; Mashanovich, G.Z.; Chong, H.M.H. Scattering Loss Estimation Using 2-D Fourier Analysis and Modeling of Sidewall Roughness on Optical Waveguides. *IEEE Photonics J.* **2013**, *5*, 6601010. [CrossRef]
2. Poulton, C.G.; Koos, C.; Fujii, M.; Pfrang, A.; Schimmel, T.; Leuthold, J.; Freude, W. Radiation Modes and Roughness Loss in High Index-Contrast Waveguides. *IEEE J. Sel. Top. Quantum Electron.* **2006**, *12*, 1306–1321. [CrossRef]
3. Horikawa, T.; Shimura, D.; Mogami, T. Low-loss silicon wire waveguides for optical integrated circuits. *MRS Commun.* **2015**, *6*, 9–15. [CrossRef]
4. Lee, K.; Lim, D.; Luan, H.; Agarwal, A.; Foresi, J.; Kimerling, L. Effect of Size and Roughness on Light Transmission in a Si/SiO2 Waveguide Experiments and Model. *Appl. Phys. Lett.* **2000**, *77*, 1617–1619. [CrossRef]
5. Lacey, J.; Payne, F. Radiation Loss From Planar Waveguides with Random Wall Imperfections. *IEE Proc.* **1990**, *137*, 282–288. [CrossRef]
6. Barwicz, T.; Haus, H.A. Three-Dimensional Analysis of Scattering Losses Due to Sidewall Roughness in Microphotonic Waveguides. *J. Light. Technol.* **2005**, *23*, 2719–2732. [CrossRef]
7. Guiana, B.; Zadehgol, A. FDTD Simulation of Stochastic Scattering Loss Due to Surface Roughness in Optical Interconnects. In Proceedings of the 2022 United States National Committee of URSI National Radio Science Meeting (USNC-URSI NRSM), Boulder, CO, USA, 4–8 January 2022; pp. 1–2.
8. Payne, F.; Lacey, J. A Theoretical Analysis of Scattering Loss from Planar Optical Waveguides. *Opt. Quantum Electron.* **1994**, *26*, 977–986. [CrossRef]
9. Guiana, B.; Zadehgol, A. Stochastic FDTD Modeling of Propagation Loss due to Random Surface Roughness in Sidewalls of Optical Interconnects. In Proceedings of the United States National Committee of URSI National Radio Science Meeting (USNC-URSI NRSM), Virtual, 4–9 January 2021; pp. 266–267.
10. Zadehgol, A. Complex s-Plane Modeling and 2D Characterization of the Stochastic Scattering Loss in Symmetric Dielectric Slab Waveguides Exhibiting Ergodic Surface-Roughness with an Exponential Autocorrelation Function. *IEEE Access* **2021**, *9*, 92326–92344. [CrossRef]
11. Kuznetsov, M.; Haus, H.A. Radiation Loss in Dielectric Waveguide Structures by the Volume Current Method. *IEEE J. Quantum Electron.* **1983**, *QE-19*, 1505–1514. [CrossRef]
12. Python Programming Language. Available online: https://www.python.org/ (accessed on 20 October 2021).
13. Zadehgol, A. SHF: SMALL: A Novel Algorithm for Automated Synthesis of Passive, Causal, and Stable Models for Optical Interconnects. National Science Foundation (NSF) Award #1816542. Proposal submitted on 11/15/2017. Grant period: 10/1/2018-9/30/2021. Available online: https://nsf.gov/awardsearch/showAward?AWD_ID=1816542&HistoricalAwards=false (accessed on 20 October 2021).
14. Guiana, B. Optical Interconnect Designer Tool (OIDT). Available online: https://github.com/bmguiana/OIDT (accessed on 23 November 2021).
15. GNU General Public License v3.0. Available online: https://www.gnu.org/licenses/gpl-3.0.en.html (accessed on 20 October 2021).
16. Yee, K.S. Numerical Solution of Initial Boundary Value Problems Involving Maxwells Equations in Isotropic Media. *IEEE Trans. Antennas Propag.* **1966**, *Ap14*, 302.
17. Taflove, A.; Hagness, S.C. *Computational Electrodynamics the Finite-Difference Time-Domain Method*, 3rd ed.; Artech House Inc.: Norwood, MA, USA, 2005.
18. Roden, J.A.; Gedney, S.D. Convolution Pml (CPML) an Efficient Fdtd Implementation of the Cfs—Pml for Arbitrary Media. *Microw. Opt. Technol. Lett.* **2000**, *27*, 334–339. [CrossRef]
19. Guiana, B.; Zadehgol, A. S-Parameter Extraction Methodology in FDTD for Nano-Scale Optical Interconnects. In Proceedings of the 15th International Conference on Advanced Technologies, Systems and Services in Telecommunications, Nis, Serbia, 20–22 October 2021; pp. 1–4.
20. Zadehgol, A. Deterministic Reduced-Order Macromodels for Computing the Broadband Radiation-Field Pattern of Antenna Arrays in FDTD. *IEEE Trans. Antennas Propag.* **2016**, *64*, 2418–2430. [CrossRef]
21. Cooley, J.W.; Tukey, J.W. An Algorithm for the Machine Calculation of Complex Fourier Series. *Math. Comput.* **1965**, *19*, 297–301. [CrossRef]
22. Pyspeckle. Available online: https://pyspeckle2.readthedocs.io/en/latest/# (accessed on 15 October 2021).

23. Deserno, M. *How to Generate Exponentially Correlated Gaussian Random Numbers*; Department of Chemistry and Biochemistry UCLA: Los Angeles, CA, USA, 2002.
24. Smith, S.W. *The Scientist and Engineer's Guide to Digital Signal Processing*; California Technical Publishing: San Diego, CA, USA, 1997; Chapter 15, pp. 277–284.
25. Balanis, C.A. *Advanced Engineering Electromagnetics*, 2nd ed.; Wiley: Hoboken, NJ, USA, 2012; pp. 270–275. 428.
26. Pozar, D.M. *Microwave Electronics*, 4th ed.; Wiley: Hoboken, NJ, USA, 2012; pp. 178–188.
27. Brown, J.W.; Churchill, R.V. *Complex Variables and Applications*, 8th ed.; McGraw-Hill: New York, NY, USA, 2009; pp. 93–99.
28. Zadehgol, A.; Cangellaris, A.C. Isotropic Spatial Filters for Suppression of Spurious Noise Waves in Sub-Gridded FDTD Simulation. *IEEE Trans. Antennas Propag.* **2011**, *59*, 3272–3279. [CrossRef]

Article

Calculation of Magnetic Flux Density Harmonics in the Vicinity of Overhead Lines

Adnan Mujezinović *, Emir Turajlić, Ajdin Alihodžić, Maja Muftić Dedović and Nedis Dautbašić

Faculty of Electrical Engineering, University of Sarajevo, 71000 Sarajevo, Bosnia and Herzegovina;
eturajlic@etf.unsa.ba (E.T.); aalihodzic1@etf.unsa.ba (A.A.); maja.muftic-dedovic@etf.unsa.ba (M.M.D.);
nd15231@etf.unsa.ba (N.D.)
* Correspondence: adnan.mujezinovic@etf.unsa.ba

Abstract: This paper considers the method for the calculation of magnetic flux density in the vicinity of overhead distribution lines which takes into account the higher current harmonics. This method is based on the Biot–Savart law and the complex image method. The considered method calculates the values of the magnetic flux density for each harmonic component of the current separately at all points of interest (usually lateral profile). In this way, it is possible to determine the contributions of individual harmonic components of the current intensity to the total value of magnetic flux density. Based on the contributions of individual harmonic components, the total (resultant) value of the magnetic flux density at points of interest is determined. Validation of the computational method is carried out by comparison of the results obtained by the considered calculation method with measurement results. Furthermore, the application of the calculation method was demonstrated by calculating magnetic flux density harmonics in the vicinity of two overhead distribution lines of typical phase conductor arrangements.

Keywords: Biot–Savart (BS) law based method; current intensity harmonics; magnetic flux density harmonics; overhead lines

Citation: Mujezinović, A.; Turajlić, E.;
Alihodžić, A.; Dedović, M.M.;
Dautbašić, N. Calculation of
Magnetic Flux Density Harmonics in
the Vicinity of Overhead Lines.
Electronics **2022**, *11*, 512. https://
doi.org/10.3390/electronics11040512

Academic Editors: Giulio Antonini,
Daniele Romano and Luigi Lombardi

Received: 30 December 2021
Accepted: 5 February 2022
Published: 9 February 2022

Publisher's Note: MDPI stays neutral
with regard to jurisdictional claims in
published maps and institutional affiliations.

1. Introduction

Advances in technology and increased energy requirements have brought with them a significant increase in the number of nonlinear loads, as well as the use of converter-based distributed generation units. Nowadays, a large number of such units are connected to the networks. As previously stated, this brings into the networks the problem of harmonic pollution [1,2].

The presence of a large number of nonlinear devices in the network causes harmonic pollution to become one of the most common power quality problems. Harmonic currents cause numerous detrimental side effects (heating, accelerated aging and capacity reduction), primarily on lines and transformers, but also on other equipment in the distribution network. They can also cause unwanted activation of protection devices, malfunctions of sensitive devices, interference in communication systems, and the reduction of useful torque of induction motors [3–5].

Due to the negative effects that harmonics have on the electrical equipment, which results in financial losses, it is important to limit harmonic currents injection [6]. For this purpose, international and national organizations defined the limits for each harmonic amplitudes, in relation to the fundamental one, as well as the limits of the total harmonic distortion (THD) [7–9]. "IEEE Recommended Practice and Requirements for Harmonic Control in Electric Power Systems" limits the harmonic distortion at the point of common coupling, and defines the limit values for up to the 50th harmonic. This standard prescribes different harmonic limits depending on the system rated voltage [7]. International Electrotechnical Commission published series of standards "IEC 61000-3:2022 SER-Electromagnetic compatibility" where different standards treat harmonic emissions in public low-voltage systems and the connection of distorting installations in medium, high and extra high voltage systems [9].

In order to meet these limitations, various methods are being developed to mitigate harmonic emission. Harmonic emission mitigation methods are classified into three categories: passive, active, and hybrid methods. Choosing the most appropriate method for mitigating harmonic emission is a demanding process, and it should be borne in mind that some of the methods are highly dependent on network conditions, while others can cause resonant phenomena [10,11].

Nonlinear loads have a tendency to change their input impedance over the time and this has a significant impact on their current harmonic emission. Thus, the network harmonic impedance can be used to characterize the impact of individual nonlinear devices on harmonic currents emission in the grid [12], which can be useful for proper harmonic limiting method selection.

Investigations of possible adverse health issues associated with exposure to magnetic fields is of great research interest in recent decades. Numerous studies are conducted in order to find the exact mechanism of the impact of these phenomena on humans, as well as to establish consequences of that exposure [13–17]. Based on the experimental research and the statistical analyses of the relationship between the magnetic field exposure and the occurrences of severe diseases, the International Commission on Non-Ionizing Radiation Protection (ICNIRP) [18] published guidelines for limiting exposure to time-varying electric and magnetic fields in the frequency range (1 Hz–100 kHz). These guidelines relate to instances when the fields are caused by pure sine currents and voltages, but also instances when their harmonic distortion is present. The guidelines cover all magnetic flux density harmonics which may occur in power system facilities. Since harmonic standards cover limit values up to the 50th harmonic, which for a system with fundamental frequency of 50 Hz corresponds to a frequency 2500 Hz, or 3000 Hz for the systems with a fundamental frequency of 60 Hz, it is interesting to note exposure limit values for that frequency range. For occupational exposure to time varying magnetic fields ICNIRP limit values are 1×10^{-3} T in a frequency range from 25 Hz to 300 Hz, and $0.3/f$ T for a frequency range from 300 Hz to 3 kHz, where f is the frequency of the analysed field. For general public exposure, the limit values for various frequency ranges are defined in a slightly different way. For a frequency range from 50 Hz to 400 Hz the limit value is 2×10^{-4} T and for a frequency range from 400 Hz to 3 kHz, the limit value is defined as $8 \times 10^{-2}/f$ T. When it comes to simultaneous exposure to multiple frequency fields, which is the case in the vicinity of power facilities with currents polluted by harmonics, specific criterion regarding reference levels should be applied [18]. According to these criteria, the influence of fields of all frequencies is summed up and the exposure limit value can be violated even if the fields at all frequencies individually meet the limit values.

It is important to perform magnetic flux density determination in the vicinity of overhead lines, as one of the strongest field sources. It is often assumed that if the magnetic flux density of the fundamental harmonic satisfies the prescribed reference levels than the harmonic magnetic flux density also satisfies them. However, this does not have to be the case in situations where currents with a large presence of higher harmonics flow through overhead lines [19,20].

In this paper, a method for magnetic flux density harmonics calculation in the vicinity of overhead lines is considered. The method is validated by comparing the results obtained by the considered method to the measurement results. Furthermore, the considered method is applied to calculate magnetic flux density harmonics in the case of a currently operational medium voltage overhead lines.

The remainder of this paper is organized as follows. Section 2 provides a detailed description of the considered calculation method for the calculation of magnetic flux density harmonics. In Section 3, a comparison of the results obtained by considered calculation method with the measurement results is given. Section 4 considers the application of the considered calculation method on a case of two overhead distribution lines of a typical configuration of phase conductors. Section 5 concludes the paper.

2. Calculation Method

In order to analyse magnetic flux density harmonics it is necessary to consider the exact phase current conditions. In overhead lines with the presence of higher current harmonics, phase current waveforms have the general form:

$$i_a(t) = \sqrt{2} \sum_{h=1}^{N} I_a^{(h)} sin(h\omega t + \theta_a^{(h)})$$ (1)

$$i_b(t) = \sqrt{2} \sum_{h=1}^{N} I_b^{(h)} sin(h\omega t + \theta_b^{(h)})$$ (2)

$$i_c(t) = \sqrt{2} \sum_{h=1}^{N} I_c^{(h)} sin(h\omega t + \theta_c^{(h)}),$$ (3)

where: $i_a(t)$, $i_b(t)$, $i_c(t)$ are phase current waveforms, N is the total number of considered harmonics, h denotes the harmonic under a consideration, $I_a^{(h)}$, $I_b^{(h)}$, $I_c^{(h)}$ are root mean square (RMS) values of phase current intensities of h-th harmonic, ω is fundamental harmonic angular frequency, and θ_a^h, θ_b^h, θ_c^h are h-th harmonic phase angles associated with the phase currents.

It is important to note that not all current harmonics behave in the same manner as the fundamental harmonic. According to symmetrical component theory, current harmonics can be categorized into three different sequences: direct, inverse, and zero. The phase angles for direct, inverse and zero sequence harmonics are defined by the Equations (4)–(6), respectively [19,20]:

$$\begin{bmatrix} \theta_a^{(h)} \\ \theta_b^{(h)} \\ \theta_c^{(h)} \end{bmatrix} = \begin{bmatrix} 0 \\ e^{-j\frac{2\pi}{3}} \\ e^{j\frac{2\pi}{3}} \end{bmatrix} (h = 3 \cdot k + 1, k = 0, 1, 2, ...)$$ (4)

$$\begin{bmatrix} \theta_a^{(h)} \\ \theta_b^{(h)} \\ \theta_c^{(h)} \end{bmatrix} = \begin{bmatrix} 0 \\ e^{j\frac{2\pi}{3}} \\ e^{-j\frac{2\pi}{3}} \end{bmatrix} (h = 3 \cdot k + 2, k = 0, 1, 2, ...)$$ (5)

$$\begin{bmatrix} \theta_a^{(h)} \\ \theta_b^{(h)} \\ \theta_c^{(h)} \end{bmatrix} = \begin{bmatrix} 0 \\ 0 \\ 0 \end{bmatrix} (h = 3 \cdot k, k = 1, 2, 3, ...).$$ (6)

For each considered calculation point in the vicinity of the overhead line, magnetic flux density can be calculated by applying the BS law based method [21]. This method considers every phase conductor as one current point source. Based on the equations for the calculation of magnetic flux density for fundamental harmonics [22], derived from the BS law based method, magnetic flux density for harmonics of a higher order can be calculated. The magnetic flux density for the entire lateral profile is calculated separately for each current harmonic. For the h-th order harmonic, the magnetic flux density phasor spatial components at an arbitrary calculation point $P(x, y)$ in the vicinity of the overhead distribution line can be calculated using the following equations:

$$\underline{B}_x^{(h)}(x,y) = \frac{\mu_0}{2\pi} \cdot \sum_{i=1}^{n} \left(-\frac{y - y_i}{r_i^2} + \frac{y + y_i + \alpha^{(h)}}{r_i'^2} \right) \cdot \underline{I}_i^{(h)}$$ (7)

$$\underline{B}_y^{(h)}(x,y) = \frac{\mu_0}{2\pi} \cdot \sum_{i=1}^{n} \left(\frac{x - x_i}{r_i^2} - \frac{x - x_i}{r_i'^2} \right) \cdot \underline{I}_i^{(h)},$$ (8)

where $\underline{B}_x^{(h)}(x,y)$ and $\underline{B}_y^{(h)}(x,y)$ are x and y vector components of the h-th harmonic magnetic flux density phasor, μ_0 is the permeability of the air, n is the total number of phase conductors, $\underline{I}_i^{(h)}$ is the current intensity of h-th harmonic of i-th phase conductor, (x,y) are coordinates of an arbitrary calculation point, (x_i,y_i) are coordinates of the i-th phase conductor, $\underline{\alpha}^{(h)}$ is the complex depth, r_i is the shortest distance between the i-th phase conductor and calculation point, r_i' is the shortest distance between complex image of i-th phase conductor and calculation point.

The effect of ground surface is taken into account by using the complex image method. The complex depth is frequency dependent, therefore it can be calculated using the following equation [23,24]:

$$\underline{\alpha}^{(h)} = \frac{2}{\sqrt{-j \cdot \omega^{(h)} \cdot \mu_0 \cdot \left(\sigma_{soil} - j \cdot \omega^{(h)} \cdot \varepsilon_{soil}\right)}}, \tag{9}$$

where $\omega^{(h)}$ is the h-th harmonic angular frequency, σ_{soil} is soil conductivity, ε_{soil} is dielectric permittivity of the soil.

The total magnetic flux density spatial vector components at an arbitrary calculation point can be obtained by applying the principle of superposition to include the contributions of each considered harmonic, as in:

$$B_x^{tot}(x,y) = \sqrt{\sum_{h=1}^{N} \left| \underline{B}_x^{(h)}(x,y) \right|^2} \tag{10}$$

$$B_y^{tot}(x,y) = \sqrt{\sum_{h=1}^{N} \left| \underline{B}_y^{(h)}(x,y) \right|^2}, \tag{11}$$

where $B_x^{tot}(x,y)$ and $B_y^{tot}(x,y)$ are magnetic flux density vector spatial components RMS values.

The resultant value of the total magnetic flux density with the presence of higher harmonics taken into account at some arbitrary point is defined as [21]:

$$B^{tot}(x,y) = \sqrt{\left| B_x^{tot}(x,y) \right|^2 + \left| B_y^{tot}(x,y) \right|^2}, \tag{12}$$

where $B^{tot}(x,y)$ is the resultant value of the magnetic flux density at an arbitrary point with coordinates (x,y).

THD of the magnetic flux density can be calculated using the following equation [25]:

$$THD_B(x,y) = \frac{\sqrt{\sum_{h=2}^{N} B^{(h)}(x,y)^2}}{B^{(1)}(x,y)}. \tag{13}$$

The RMS value of magnetic flux density for each considered harmonic, including the fundamental one, can be calculated from the following equation:

$$B^{(h)}(x,y) = \sqrt{\left| \underline{B}_x^{(h)}(x,y) \right|^2 + \left| \underline{B}_y^{(h)}(x,y) \right|^2}, \tag{14}$$

where $B^{(h)}(x,y)$ is the RMS value of the h-th magnetic flux density at an arbitrary point with coordinates (x,y).

3. Calculation Method Validation

In order to validate the proposed calculation method, a comparison between calculated and measured values is presented in this section. Two cases have been analysed. The first considered case is a simple low voltage three-phase horizontal conductor arrangement with previously published results. The second considered case corresponds to a three-phase distribution overhead line of 35kV rated voltage with horizontal phase conductor arrangement.

3.1. Low-Voltage Experimental Case

The method for the calculation of magnetic flux density harmonics in the vicinity of overhead lines is validated by the comparison of calculated magnetic flux density with measurement results given in the literature [26]. The measurements correspond to the three-phase configuration where conductors are placed at the same height, as shown in Figure 1a. Magnetic flux density calculation and measurements are performed at ground level over the considered lateral profile in increments of 25 cm. Two cases are examined, without the presence of higher current harmonics, and with the presence of third current harmonic. The phase current intensity RMS value during the measurements is 4.58 A at frequency 16.666 Hz, and the third harmonic current intensity is 0.7633 A at frequency 50 Hz [26]. For both cases, the comparison of magnetic flux density measurement results and the obtained calculation results for the considered test case are given in Figure 1b.

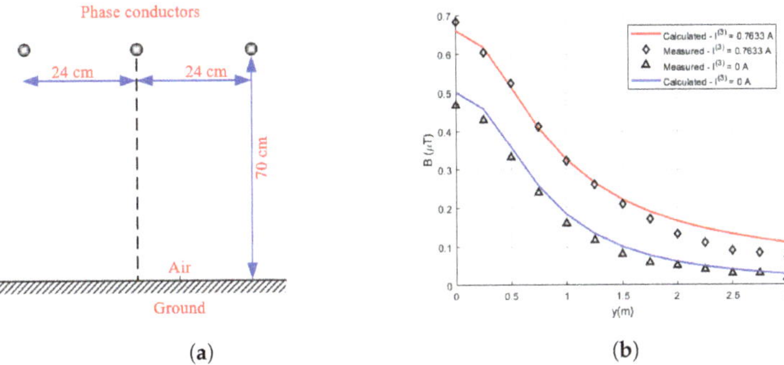

(**a**)   (**b**)

Figure 1. Considered validation case study. (**a**) Three-phase experimental configuration [26]. (**b**) Comparison of calculated and measured results.

From Figure 1b, it can be noted that there exists a high level of agreement between the measurement and the calculation magnetic flux density results in both considered cases; in the absence and in the presence of a third current harmonic in the system. Although magnetic flux density measurements were originally presented in Gauss [26], in this paper they are converted to SI system units. Measurements were extracted from the paper [26] using the WebPlotDigitizer tool [27], which may be the cause of certain deviations from the originally presented measurements.

In order to determine the level of agreement between the calculated and measured values, in both cases, mean square error (MSE) metric is used. In the case when there is no presence of higher harmonics, the mean square error is 4.08×10^{-4} $(\mu T)^2$, while in the case when there is a third harmonic in the considered system, the MSE value amounts to 6.62×10^{-4} $(\mu T)^2$.

There are certain deviations between the measurement and calculation results regardless of whether there is the presence of higher harmonics or not. It is interesting to note that, in the significant number of considered points observed deviations are larger in case without presence of third harmonic. Taking this into account it can be considered that the proposed model adequately takes into account the presence of higher harmonics. Deviations between measured and calculated values in both cases can be caused by low current intensity, which causes low magnetic flux density values, which are more difficult to register. Furthermore, although measurements are intended to be performed at floor level, it is difficult to set the sensor exactly at that level and due to the small height at which conductors are placed, this can significantly affect measured values. The small distances between the conductors and between the points where the magnetic flux density measurement are performed, can also have an impact on measured values.

3.2. Overhead Distribution Line Case

The proposed calculation model is also validated in the case of the overhead distribution line of 35 kV rated voltage. The phase conductor configuration of considered distribution line is shown in Figure 2a. The height of the phase conductors was determined by measurements using Suparule model 600 height meter, while the lateral distance between the phase conductors was taken from the technical specifications obtained from the local distribution company. The magnetic flux density has been measured by the 3D probe Narda-ELT 400 whose frequency range is from 1 Hz up to 400 kHz. Measurements are performed at the height of 1 m above ground surface. Measurements were performed from the distribution line axis to a distance of 20 m in one direction with increments of 1 m. Simultaneously with the measurement of the magnetic flux density, the current intensity was also measured. The measurement of the current intensity and current harmonic spectrum was conducted by power quality analyser EPPE W8. These measurements were carried out in the distribution transformer station (35/10 kV) which is supplied by the considered overhead distribution line. The power quality analyser was connected to the secondary windings of current measuring transformers that are placed on the lower voltage terminals (10 kV terminals) of the distribution transformer. The transmission ratio of the current measuring transformer is 600/5 (A). Therefore, the measured values of current intensity were corrected by the transmission ratios of the current measuring transformer and the distribution transformer. The value of the fundamental harmonic intensity was 218.64 A at frequency 50 Hz. The registered current harmonic content is given in Figure 2b.

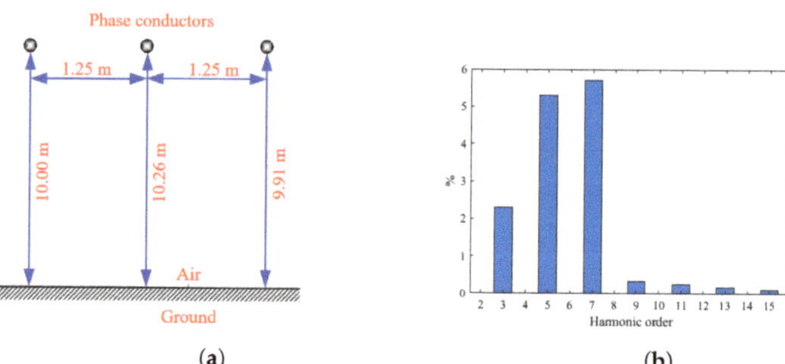

Figure 2. Considered 35 kV horizontal overhead distribution line case. (**a**) 35 kV overhead distribution line configuration. (**b**) Current harmonic spectrum.

A comparison between the measured and calculated magnetic flux density values is presented in Figure 3. It can be noted that the results of measurements and the calculations of magnetic flux density closely match. For this case, the calculated value of the mean square error equals 2.24×10^{-3} $(\mu T)^2$.

The observed deviations between the measurements and calculation results can be caused by numerous factors such as current intensity and current harmonic content variation during the measurement of the magnetic flux density, measurement transformer error and uneven terrain on location where magnetic flux density was measured.

The results of two analysed cases demonstrate that the considered calculation method can be used for magnetic flux density analysis in the vicinity of overhead lines with the presence of higher current harmonics.

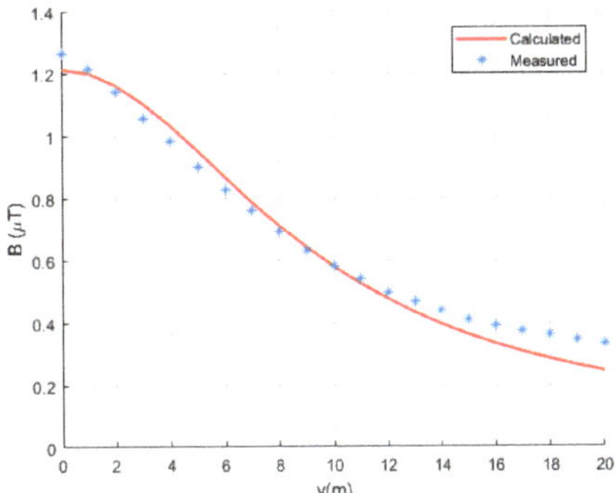

Figure 3. Comparison of calculated and measured values for analysed 35 kV overhead distribution line.

4. Case Studies

The considered calculation method has been applied to perform a magnetic flux density harmonics calculation in the vicinity of two typical overhead distribution lines. The first case corresponds to a 10 kV overhead distribution line of horizontal phase conductor configuration and the second case corresponds to a 35 kV overhead distribution line. These overhead distribution line configurations are selected based on their widespread use, large human population proximity, and the presence of higher current harmonics in the medium voltage networks, due to a high number of nonlinear loads and distributed generation.

4.1. Overhead Distribution Line of Horizontal Configuration

The first considered overhead distribution line configuration is shown in Figure 4a. Calculations are performed for the rated current of phase conductors and the usual presence of higher current harmonics. For the 10 kV rated voltage overhead line, the aluminium steel conductor with ampacity of 170 A is typically used. Calculations are performed for the converter-based distributed generator typical harmonic spectrum as in Table 1 [28,29].

Table 1. Considered current harmonic spectrum.

Harmonic Order	Magnitude (%)	Phase Angle (Rad)
1	100	0
5	20	0
7	14.3	0
11	9.1	0
13	7.7	0
17	5.9	0
19	5.3	0
23	4.3	0
25	4	0
29	3.4	0
31	3.2	0

The main goal of magnetic flux density calculation is to determine whether or not the values of magnetic flux density generated by the overhead distribution line exceed the prescribed values. Thus, in Figure 4b, the maximum calculated values of magnetic flux density harmonics are shown together with the ICNIRP reference levels for exposure to time varying magnetic fields [18]. Calculations are performed over the lateral profile from −20 m to +20 m from distribution line axis, at a height of 1 m above ground surface. Calculations are performed for the assumed minimum conductor heights, where the highest magnetic flux density values are expected [30].

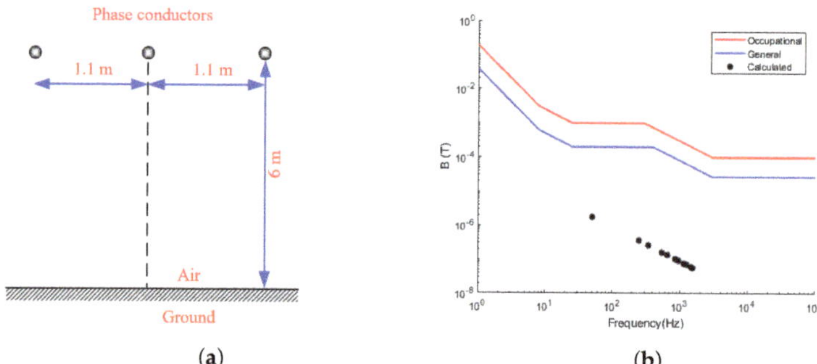

(**a**)

(**b**)

Figure 4. Considered 10 kV overhead distribution line case. (**a**) 10 kV overhead distribution line configuration. (**b**) Comparison with ICNIRP reference values.

From Figure 4b it can be noted that all magnetic flux density harmonics, in the considered case study, satisfy prescribed limit values for both general and occupational exposure. In addition, ICNIRP proposes criteria for simultaneous exposure to multiple frequency fields [18]. These criteria are originally defined for magnetic field strength, but in this paper they are adjusted for the magnetic flux density. Taking into account the relationship between the magnetic field strength and the magnetic flux density, the criteria can be defined by the following equation:

$$\sum_{j=1\,\text{Hz}}^{10\,\text{MHz}} \frac{B_j}{B_{R,j}} \leq 1,\tag{15}$$

where B_j is magnetic flux density at frequency j, and $B_{R,j}$ is magnetic flux density reference level at frequency j for professional or general public exposure.

Taking into account magnetic flux densities of all present harmonics, the sum from Equation (15) corresponds to a value of 4.8×10^{-3}, which is significantly less than 1. The general public exposure reference values are used for calculation. Considering that occupational exposure reference values are larger, it can be concluded that for this test case exposure limits are satisfied in all conditions.

Taking into account that a typical overhead distribution line is analysed under the assumption that phase current intensity is equal to the conductor ampacity, information that even in the most difficult load conditions will not violate the prescribed reference values is very significant. This is especially relevant for densely populated areas, where distribution lines are widespread.

Figure 5a presents the magnetic flux density distributions associated with the fundamental harmonic, the total magnetic flux density where contributions of the entire analyzed harmonic spectrum are taken into account, and the magnetic flux density where only a limited harmonic spectrum is considered (only those harmonics with values higher than 5% of fundamental harmonic are considered). Figure 5b shows the distribution of magnetic flux density higher order harmonics over the considered lateral profile.

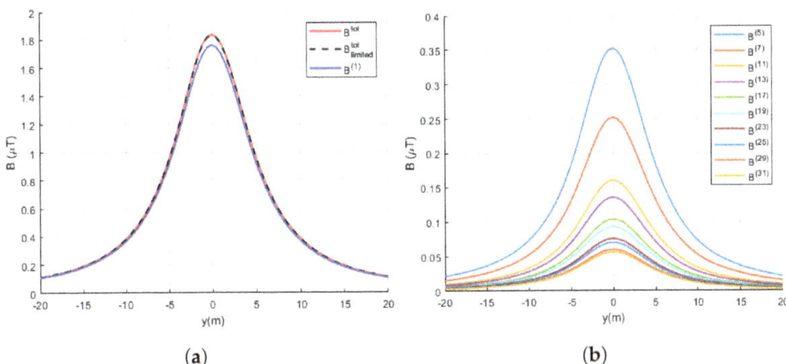

(a)   (b)

Figure 5. Magnetic flux density distribution for the analysed 10 kV overhead distribution line. (**a**) Fundamental harmonic and total magnetic flux density. (**b**) Higher harmonics magnetic flux density.

It can be noted from Figure 5 that all considered magnetic flux density harmonics as well as the total magnetic flux density have a maximum value at the distribution line axis. The maximum calculated values of magnetic flux density in Figure 5a are 1.84 µT for total magnetic flux density with entire harmonic spectrum, 1.83 µT for limited harmonic spectrum and 1.76 µT for fundamental harmonic magnetic flux density. The relative error between the total magnetic flux densities, when the entire harmonic spectrum and limited harmonic spectrum were taken into account, is −0.54%. On the other hand, the highest relative error between the magnetic flux densities corresponding to a case when the entire harmonic spectrum is considered, and a case when only the fundamental harmonic is taken into account, is equal to −4.35%. The observed difference between these three values is due to the presence of higher current harmonics which cause higher magnetic flux density harmonics.

Fundamental harmonic magnetic flux density, total magnetic flux density and magnetic flux density limited to harmonics higher than 5% over one period are shown in Figure 6. Magnetic flux density waveforms correspond to the point directly under the central phase conductor, where magnetic flux density reaches the maximum value for all considered harmonics. The total magnetic flux density waveforms when higher harmonics presence was taken into account are obtained by superimposing the waveforms of all considered magnetic flux density harmonics.

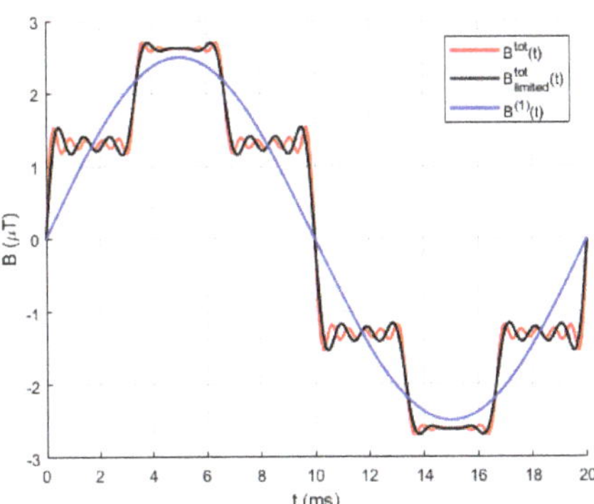

Figure 6. Magnetic flux density waveform for analysed 10 kV overhead distribution line.

As it can be noted from Figure 6 fundamental harmonic magnetic flux density have the sine waveform, while other two magnetic flux densities have distorted waveforms due to the presence of higher current harmonics. Also, it can be noted that magnetic flux densities for cases of entire and limited harmonic spectrum have similar waveforms with minor deviations.

4.2. Overhead Distribution Line of 35 kV Rated Voltage

Phase conductor arrangement of the second analysed overhead distribution line is presented in Figure 7a. The calculations are performed for the rated current of phase conductors and in the presence of higher current harmonics. For the analysed 35 kV rated voltage distribution line, the current intensity of fundamental harmonic of 300 A is assumed. The harmonic spectrums used in this case are the same as in the previous case. In Figure 7b, the maximum calculated values of magnetic flux density harmonics are shown together with the ICNIRP reference levels for exposure to time varying magnetic fields [18]. Calculations are performed over the lateral profile from −20 m to +20 m from distribution line axis, with increments of 1 m. Calculations are performed at a height of 1 m above ground surface.

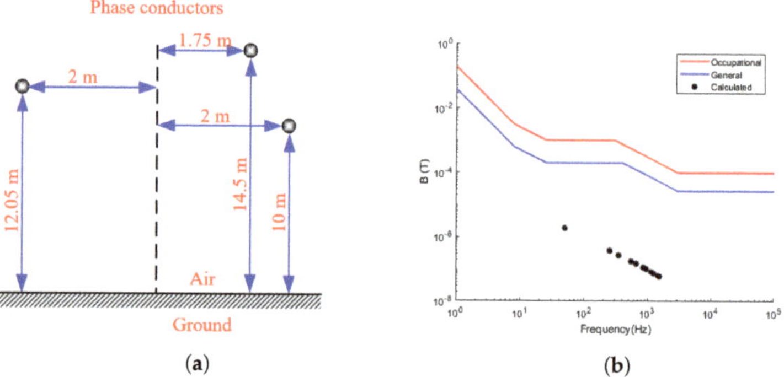

(**a**) (**b**)

Figure 7. Considered 35 kV overhead distribution line case. (**a**) 35 kV overhead distribution line configuration. (**b**) Comparison with ICNIRP reference values.

In this case, as in the previous one, all magnetic flux density harmonics are lower than both prescribed limit values for the general and occupational exposure. Applying the ICNIRP criteria for simultaneous exposure to multiple frequency magnetic fields, the obtained sum for stricter limits, defined in Equation (15) amounts to 5.2×10^{-3}, which means that also in this considered case, exposure restrictions are satisfied.

The comparison of magnetic flux density distribution for fundamental harmonics as well as the total magnetic flux density, where contributions of the entire analysed harmonic spectrum and limited harmonic spectrum are considered, are shown in Figure 8a. The magnetic flux density of higher order harmonics over the considered lateral profile is shown in Figure 8b.

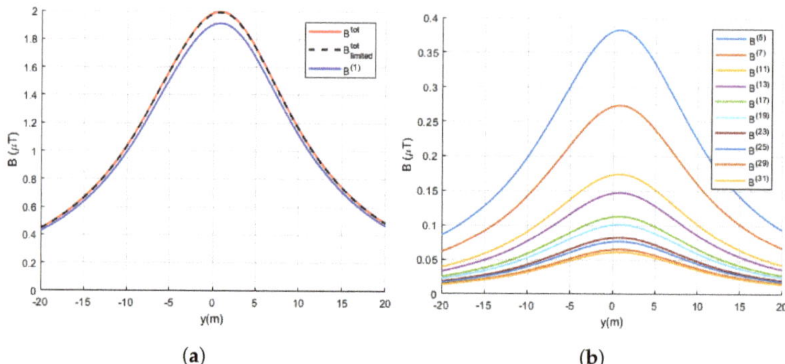

(**a**) (**b**)

Figure 8. Magnetic flux density distribution for analysed 35 kV overhead distribution line. (**a**) Fundamental harmonic and total magnetic flux density. (**b**) Higher harmonics magnetic flux density.

The maximum calculated value of the fundamental harmonic magnetic flux density shown in Figure 8a is 1.91 μT. On the other hand, the maximum values of the total magnetic field density for case of entire and limited harmonic spectrum are 2.00 μT and 1.99 μT, respectively. The highest relative error between total magnetic flux densities when entire harmonic spectrum and limited harmonic spectrum were taken into account is −0.5%. The highest relative error between the magnetic flux densities corresponding to a case when entire harmonic spectrum is considered, and a case when only fundamental harmonic is taken into account, is equal to −4.5%. It is noticeable that in this case, the higher current harmonics have a significant impact on the calculated magnetic flux density values, as in the previous case.

Fundamental harmonic magnetic flux density waveform, magnetic flux density waveform obtained by taking into account entire harmonic spectrum and magnetic flux density taking into account limited harmonic spectrum for the analysed 35 kV overhead distribution line are shown in Figure 9. The waveform correspond to a particular point in the considered lateral profile where maximum magnetic flux density values are attained.

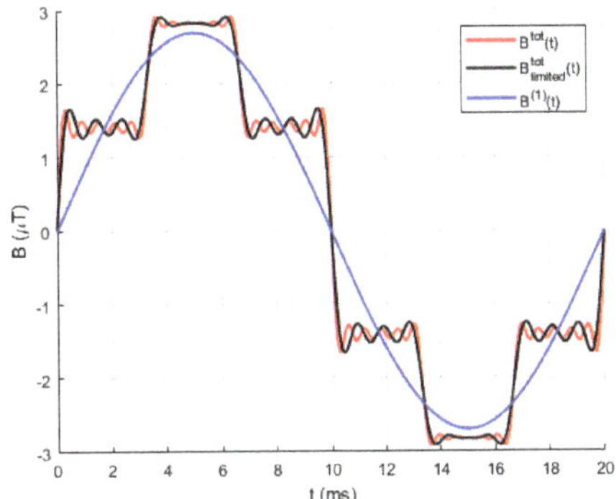

Figure 9. Magnetic flux density waveform for analysed 35 kV overhead distribution line.

Figure 9 demonstrates the distortion of magnetic flux density waveform due to the impact of the current harmonics. The total magnetic flux density waveform is dependent on the current harmonic spectrum.

Magnetic flux density THD, over the entire considered lateral profile, is 29.42% in both considered case studies and, as expected, it is equal to current intensity THD. At all analyzed points, the ratio of the magnetic flux density of the higher harmonic to the fundamental harmonic magnetic flux density is also constant and equal to the ratio of the current of the corresponding higher harmonic to the fundamental harmonic current.

The obtained results, in all analysed cases, undoubtedly show the significance of magnetic flux density harmonics in the magnetic flux density exposure analysis. The magnetic flux density harmonics are especially important when analyzing overhead distribution lines, due to the their harmonic pollution as well as population proximity.

5. Conclusions

This paper consider a method for the calculation of magnetic flux density harmonics in the vicinity of overhead lines. The considered method is based on the BS law. The validation of the calculation method is done by comparison of the results obtained by the considered calculation method with the measurement results. In addition, the considered method is applied for magnetic flux density harmonics calculation of two typical overhead distribution lines. The obtained results show the difference between the fundamental harmonic magnetic flux density and the resultant value obtained when the higher current harmonics are taken into consideration. The observed differences emphasize the significance of magnetic flux density harmonics analyses, especially in situations where the obtained magnetic flux density values are close to the prescribed values.

Results presented within the paper showed that, when calculating the magnetic flux density caused by harmonics distorted current, the magnetic flux density waveform will also be distorted. In addition, it was shown that if current harmonics with a small percentage share are neglected, this will have a slight effect on a calculated total value of the magnetic flux density. Specifically, it was shown that when the share of current harmonics is lower than 5%, they can be neglected without introducing any significant errors in the results of the total magnetic flux density calculation. Such an approximation further reduces the calculation time, since the total current harmonic spectrum does not need to be considered.

Author Contributions: A.M., E.T. and A.A. have implemented the magnetic flux density calculation method. Writing and editing, A.M., E.T., A.A., M.M.D. and N.D. All authors have read and agreed to the published version of the manuscript.

Funding: This research and publication was funded by Ministry for Science, Higher Education and Youth of Sarajevo Canton.

Institutional Review Board Statement: Not applicable.

Informed Consent Statement: Not applicable.

Data Availability Statement: Not applicable.

Conflicts of Interest: The authors declare no conflict of interest.

References

1. Barutcu, I.C.; Karatepe, E.; Boztepe, M. Impact of harmonic limits on PV penetration levels in unbalanced distribution networks considering load and irradiance uncertainty. *Int. J. Electr. Power Energy Syst.* **2020**, *118*, 105780. [CrossRef]
2. Wang, R.; Tian, J.; Wu, F.; Zhang, Z.; Liu, H. PSO/GA Combined with Charge Simulation Method for the Electric Field Under Transmission Lines in 3D Calculation Model. *Electronics* **2019**, *8*, 1140. [CrossRef]
3. Michalec, Ł.; Jasiński, M.; Sikorski, T.; Leonowicz, Z.; Jasiński, Ł.; Suresh, V. Impact of Harmonic Currents of Nonlinear Loads on Power Quality of a Low Voltage Network–Review and Case Study. *Energies* **2021**, *14*, 3665. [CrossRef]
4. Maan, J.S.; Singh, S.; Singh, A. Impact of Harmonics on Power Transformer Losses and Capacity Using Open DSS. *Int. J. Emerg. Electr. Power Syst.* **2019**, *20*, 20180349. [CrossRef]
5. Lumbreras, D.; Gálvez, E.; Collado, A.; Zaragoza, J. Trends in Power Quality, Harmonic Mitigation and Standards for Light and Heavy Industries: A Review. *Energies* **2020**, *13*, 5792. [CrossRef]
6. Sinvula, R.; Abo-Al-Ez, K.M.; Kahn, M.T. A Proposed Harmonic Monitoring System for Large Power Users Considering Harmonic Limits. *Energies* **2020**, *13*, 4507. [CrossRef]
7. *IEEE Std 519-2014 (Revision of IEEE Std 519-1992);* IEEE Recommended Practice and Requirements for Harmonic Control in Electric Power Systems. IEEE: Piscataway, NJ, USA, 2014; pp. 1–29. [CrossRef]
8. *EN 50160:2010/A3:2019 (AMENDMENT);* Voltage Characteristics of Electricity Supplied by Public Electricity Networks. British Standards Institute: London, UK, 2019.
9. *IEC 61000-3:2021;* SER—Series Electromagnetic Compatibility (EMC)—Part 3: Limit—ALL PARTS. International Electrotechnical Commission: Geneva, Switzerland, 2021.
10. Aziz, M.H.A.; Azizan, M.M.; Sauli, Z.; Yahya, M.W. A review on harmonic mitigation method for non-linear load in electrical power system. *AIP Conf. Proc.* **2021**, *2339*, 020022. [CrossRef]
11. Kazem, H.A. Harmonic Mitigation Techniques Applied to Power Distribution Networks. *Adv. Power Electron.* **2013**, *2013*, 591680. [CrossRef]
12. Chakravorty, D.; Meyer, J.; Schegner, P.; Yanchenko, S.; Schocke, M. Impact of Modern Electronic Equipment on the Assessment of Network Harmonic Impedance. *IEEE Trans. Smart Grid* **2017**, *8*, 382–390. [CrossRef]
13. Gallastegi, M.; Zabala, A.; Santa Marina, L.; Aurrekoetxea, J.; Ayerdi, M.; Ibarluzea, J.; Kromhout, H.; González, J.; Huss, A. Exposure to extremely low and intermediate-frequency magnetic and electric fields among children from the INMA-Gipuzkoa cohort. *Environ. Res.* **2017**, *157*, 190–197. [CrossRef]
14. Amoon, A.; Crespi, C.; Ahlbom, A.; Bhatnagar, M.; Bray, I.; Bunch, K.; Clavel, J.; Feychting, M.; Hémon, D.; Johansen, C.; et al. Proximity to overhead power lines and childhood leukaemia: An international pooled analysis. *Br. J. Cancer* **2018**, *119*. [CrossRef] [PubMed]
15. Crespi, C.M.; Swanson, J.; Vergara, X.P.; Kheifets, L. Childhood leukemia risk in the California Power Line Study: Magnetic fields versus distance from power lines. *Environ. Res.* **2019**, *171*, 530–535. [CrossRef] [PubMed]
16. Pedersen, C.; Johansen, C.; Schüz, J.; Olsen, J.; Raaschou-Nielsen, O. Residential exposure to extremely low-frequency magnetic fields and risk of childhood leukaemia, CNS tumour and lymphoma in Denmark. *Br. J. Cancer* **2015**, *113*, 1370–1374. [CrossRef] [PubMed]
17. Karimi, A.; Ghadiri Moghaddam, F.; Valipour, M. Insights in the biology of extremely low-frequency magnetic fields exposure on human health. *Mol. Biol. Rep.* **2020**, *47*, 5621–5633. [CrossRef]

18. Lin, J.; Saunders, R.; Schulmeister, K.; Söderberg, P.; Stuck, B.; Swerdlow, A.; Taki, M.; Veyret, B.; Ziegelberger, G.; Repacholi, M.; et al. ICNIRP Guidelines for limiting exposure to time-varying electric and magnetic fields (1 Hz to 100 kHz). *Health Phys.* **2010**, *99*, 818–836. [CrossRef]
19. Wu, T.; Xiao, B.; Liu, K.; Liu, T.; Peng, Y.; Su, Z.; Tang, P.; Lei, X. Study on overhead transmission line magnetic field harmonics of VSC-HVDC. In Proceedings of the IEEE International Conference on High Voltage Engineering and Application (ICHVE), Chengdu, China, 19–22 September 2016; pp. 1–4. [CrossRef]
20. Brandao Faria, J.; Almeida, M. Computation of transmission line magnetic field harmonics. *Int. Trans. Electr. Energy Syst.* **2007**, *17*, 512–525. [CrossRef]
21. Mujezinovic, A.; Dautbasic, N.; Dedovic, M.M. More Accurate 2D Algorithm for Magnetic Field Calculation Under Overhead Transmission Lines. In *Advanced Technologies, Systems, and Applications IV, Proceedings of the International Symposium on Innovative and Interdisciplinary Applications of Advanced Technologies (IAT 2019), Sarajevo, Bosnia and Herzegovina, 20–23 June* 2019; Springer: Cham, Switzerland, 2019.
22. Alihodzic, A.; Mujezinovic, A.; Turajlic, E. Electric and Magnetic Field Estimation Under Overhead Transmission Lines Using Artificial Neural Networks. *IEEE Access* **2021**, *9*, 105876–105891. [CrossRef]
23. Salari, J.; Mpalantinos, A.; Silva, J. Comparative Analysis of 2-D and 3-D Methods for Computing Electric and Magnetic Fields Generated by Overhead Transmission Lines. *IEEE Trans. Power Deliv.* **2009**, *24*, 338–344. [CrossRef]
24. Boteler, D.; Pirjola, R. The complex-image method for calculating the magnetic and electric fields at the surface of the Earth by the auroral electrojet. *Geophys. J. Int.* **1998**, *132*, 31–40. [CrossRef]
25. Keikko, T.; Seesvuori, R.; Valkealahti, S. Exposure to magnetic field harmonics in the vicinity of indoor distribution substations. *Health Phys.* **2006**, *91*, 574–581. [CrossRef]
26. Hejase, H.; Shaltout, A. Impact of Third Harmonic Currents on Magnetic Field Measurements from Distribution Lines. *Electr. Power Compon. Syst.* **2004**, *32*, 353–366. [CrossRef]
27. Rohatgi, A. Webplotdigitizer: Version 4.5, 2021. Available online: https://automeris.io/WebPlotDigitizer/citation.html (accessed on 29 December 2021).
28. Milovanović, M.; Radosavljević, J.; Perović, B. A backward/forward sweep power flow method for harmonic polluted radial distribution systems with distributed generation units. *Int. Trans. Electr. Energy Syst.* **2020**, *30*, e12310. [CrossRef]
29. Ulinuha, A.; Masoum, M.; Islam, S. Harmonic power flow calculations for a large power system with multiple nonlinear loads using decoupled approach. In Proceedings of the Australasian Universities Power Engineering Conference, Perth, Australia, 9–12 December 2007; pp. 1–6. [CrossRef]
30. *IEEE Std 644-2019 (Revision of IEEE Std 644-2008)*; IEEE Standard Procedures for Measurement of Power Frequency Electric and Magnetic Fields from AC Power Lines. IEEE: Piscataway, NJ, USA, 2020; pp. 1–40.

 *electronics*

Article

Compressed Complex-Valued Least Squares Support Vector Machine Regression for Modeling of the Frequency-Domain Responses of Electromagnetic Structures

Nastaran Soleimani and Riccardo Trinchero *

Department of Electronics and Telecommunications, Politecnico di Torino, 10129 Torino, Italy; nastaran.soleimani@polito.it
* Correspondence: riccardo.trinchero@polito.it

Abstract: This paper deals with the development of a Machine Learning (ML)-based regression for the construction of complex-valued surrogate models for the analysis of the frequency-domain responses of electromagnetic (EM) structures. The proposed approach relies on the combination of two-techniques: (i) the principal component analysis (PCA) and (ii) an unusual complex-valued formulation of the Least Squares Support Vector Machine (LS-SVM) regression. First, the training and test dataset is obtained from a set of parametric electromagnetic simulations. The spectra collected in the training set are compressed via the PCA by exploring the correlation among the available data. In the next step, the compressed dataset is used for the training of compact set of complex-valued surrogate models and their accuracy is evaluated on the test samples. The effectiveness and the performance of the complex-valued LS-SVM regression with three kernel functions are investigated on two application examples consisting of a serpentine delay structure with three parameters and a high-speed link with four parameters. Moreover, for the last example, the performance of the proposed approach is also compared with those provided by a real-valued multi-output feedforward Neural Network model.

Keywords: Least Squares Support Vector Machine; serpentine delay line; high-speed interconnect link; principal component analysis

Citation: Soleimani, N.; Trinchero, R. Compressed Complex-Valued Least Squares Support Vector Machine Regression for Modeling of the Frequency-Domain Responses of Electromagnetic Structures. *Electronics* **2022**, *11*, 551. https://doi.org/10.3390/electronics11040551

Academic Editor: Manuel Arrebola

Received: 29 December 2021
Accepted: 8 February 2022
Published: 11 February 2022

Publisher's Note: MDPI stays neutral with regard to jurisdictional claims in published maps and institutional affiliations.

1. Introduction

In recent decades, Machine Learning (ML) methods have been widely applied to construct accurate and fast-to-evaluate surrogate models able to reproduce the input–output behavior of electromagnetic (EM) structures as a function of deterministic and uncertain parameters [1–14]. In the above scenario, advanced data-driven and ML-based regressions, such as Polynomial Chaos Expansion [1–4], Support Vector Machine (SVM) regression [5,6], Least-Squares Support Vector Machine (LS-SVM) regression [7], Gaussian Process regression (GPR) [8], and feedforward [9–12], deep [12,13], convolutional [12] and Long Short-Term Memory (LSTM) [14] neural networks (NNs), have been successful applied to uncertainty quantification (UQ) and optimization in EM applications. The common idea is to adopt the above methods to train a regression model by using a limited number of training samples generated via a set of expensive full-wave or circuital simulations with the so-called computational model. Then, since the resulting surrogate model is known in a closed form, it can be suitably embedded within UQ and an optimization scheme as an efficient and effective "surrogate" of the expensive computational model [15].

Despite the high performance shown in realistic electronic and EM applications, most of the ML techniques have been developed for dealing with real-valued data [16–18]. However, complex-valued data are widely used in electronic applications (e.g., for AC simulations and frequency-domain analysis). The simplest strategy for extending the applicability of real-valued ML techniques to the case of complex-valued data is based on the so-called dual-channel formulation [19–21]. The underlaying idea is to recast the complex-valued problem into two uncorrelated real-valued ones, by stacking the real and imaginary part of the complex input and output values. The main advantage of the above procedure is that plain real-valued ML techniques can be directly adopted without requiring any generalization or improvement. However, such an approach completely ignores the possible correlation among the real and imaginary parts of the complex-valued output, thus leading to possible issues regarding accuracy and robustness to noise [19,22]. For the above reasons, alternative pure complex-valued formulations have been proposed for several ML techniques, such as NN [22], SVM regression [23], kernel Least Squares regression [19–21,24,25], LS-SVM regression [7] and GPR [26].

This paper investigates the performances of a pure-complex implementation of a kernel-based ML technique such as the LS-SVM regression. The proposed complex-value formulation is based on the theoretical framework developed in [19,21,24] for a plain kernel-based regression in which the regularization term is missing. Indeed, like the SVM regression, the LS-SVM regression is a powerful and well-consolidated regression technique combining the beneficial effect of the kernel and of the Tikhonov regularization [17,25,27]. Specifically, thanks to the kernel trick, the above regression allows the construction of a non-parametric model in which the number of unknowns to be estimated by the regression problem is independent from the dimensionality of the input space. Furthermore, the regularizer terms limits the overfitting issue and increases the model generalization.

The aim of this work is twofold. First, the mathematical formulation of the complex-value LS-SVM regression is developed, and the mathematical link between the pure complex formulation and the dual channel one is highlighted, where the latter is a special case of the more general complex-valued formulation. Second, the performance of compressed surrogate models constructed via the dual-channel and the pure complex-value LS-SVM regression with a complex and pseudo kernel are investigated by considering the frequency-domain responses of two test-cases consisting of a serpentine delay structure and a high-speed link, with three and four parameters, respectively. It is important to stress that, in the above applications, a compressed representation of the frequency-domain responses based on principal component analysis (PCA) [18,28–30] has been used to remove the redundant information and to reduce the number of LS-SVM-based models that need to be trained [31]. For the sake of completeness, for the second example, the results of the proposed approaches are compared with those provided by a real-valued multi-output feedforward NN model [11,12].

The paper is organized as follows. Section 2 describes the problem statement addressed during the paper, and its challenges. Section 3 provides an overview of the mathematical background of the PCA. Section 4 provides a self-contained formulation of the complex-valued LS-SVM regression with specific emphasis on the differences between the dual-channel and the pure complex implementation. In Section 5, the accuracy and the performance of the proposed complex-valued LS-SVM regression are investigated on two test cases consisting of a serpentine delay line structure and a transfer function of a high-speed interconnect link. Finally, the paper ends with the conclusions in Section 6.

2. Problem Statement and Challenges

Starting from a training set $D = \{(\mathbf{x}_i, y_i(f_k))\}_{i,k=1}^{L,K}$ and collecting the configuration of the input parameters $\mathbf{x}_i = [x_{i,1}, \ldots, x_{i,d}]^T \in \mathcal{X}$, with $\mathcal{X} \in \mathbb{C}^d$ and the corresponding output $y_i(f_k) \in \mathbb{C}$ computed by a full-computational model (i.e., $y_i(f_k) = \mathcal{M}(f_k; \mathbf{x}_i)$) for a set of values of the independent variable f_k (e.g., the frequency), our goal is to build a surrogate model $\widetilde{\mathcal{M}}$ that approximates the training data and is able to generalize well on the "unseen" data, usually known as test samples, such as:

$$y(f_k; \mathbf{x}) \approx \widetilde{\mathcal{M}}(f_k; \mathbf{x}), \tag{1}$$

For any $\mathbf{x} \in \mathcal{X}$ and for $k = 1, \ldots, K$.

Without loss of generality, we seek a data-driven regression technique able to provide an accurate and efficient approximation of the actual behavior of the computational model $\mathcal{M}(f_k; \mathbf{x}_i)$. This means that the surrogate model $\widetilde{\mathcal{M}}$ should be constructed by using a small set of training samples (i.e., L should be as small as possible), since the training samples are usually generated via a set of computational expensive simulations based on the full-computational model $\mathcal{M}$.

The above modeling problem is rather challenging. First, the regression technique should be able to work in the complex domain. Moreover, the resulting surrogate model should be able to mimic the behavior of a multi-output complex-valued function provided by the computational model (i.e., the output values $y_i(f_k)$), in which its values unavoidably depend on both the configuration of the input parameters $\mathbf{x}_i$ and on the variable f_k.

A possible modeling scheme consists of considering the free variable f_k as an extra input parameter. This means that we are seeking a "single" advanced model able to represent in a closed-form the impact of both the parameter $\mathbf{x}$ and the frequency f_k on the system output. As an example, such a model can be obtained via a plain or recurrent NN [12,13]. However, due to the complexity of the problem at hand, the resulting neural network structure usually requires a large number of neurons and several hidden layers. Moreover, despite their flexibility, the training of NN-based structures requires the solution of a non-convex optimization leading to training issues (i.e., expensive training time and/or huge number of training samples [13,32]).

In order to partially overcome the above issues, we can think of using a single-output regression trained via the solution of a convex optimization problem, such as the plain kernel Least Squares regression [21], the LS-SVM regression [33], or the SVM regression [34], and build one model for each frequency point. In this way, the model generation requires less training samples; unfortunately, however, the overall model requires training K single-output surrogates, thus making the training process extremely expensive and cumbersome when a large number of frequency points are considered.

A data compression strategy can be seen as a good compromise between the two modeling schemes depicted in the previous section. As an example, PCA [18,28–30] allows the extraction of the inherent correlation existing among several realizations of output data samples at different frequency points, thus leading to a compressed representation of the frequency spectra. In such a scenario, the number of actual single-output models required to represent the data can be heavily reduced.

3. PCA Compression

This section briefly presents the mathematical background of the PCA [18,28–30]. Let us consider the output dataset $\{y_i(f_k)\}_{i,k=1}^{L,K}$ that collects L spectra computed for different configurations of the input parameters (i.e., the number of training outputs), each having K frequency points. The above dataset can be recast as a $K \times L$ matrix $\mathbf{Y}$, such that the element $Y_{i,k} = y_i(f_k)$. For normalization purposes, a zero-mean matrix is considered:

$$\widetilde{\mathbf{Y}} = \mathbf{Y} - \mu \, , \tag{2}$$

where μ is the mean value of $\mathbf{Y}$, calculated row-wise and subtracted column-wise

The new rectangular matrix $\widetilde{\mathbf{Y}} \in \mathbb{C}^{K \times L}$ can be represented by means of the compact singular value decompositions (SVD) [28,30]:

$$\widetilde{\mathbf{Y}} = \mathbf{U}\mathbf{S}\mathbf{V}^{\mathrm{H}}, \tag{3}$$

where, assuming that there are L columns of $\mathbf{U}$ and $\mathbf{V}$ associated with non-zero singular values, $\mathbf{U} = [\mathbf{u}_1, \dots \mathbf{u_L}] \in \mathbb{C}^{K \times L}$ and $\mathbf{V} = [\boldsymbol{v_1}, \dots, \boldsymbol{v_L}] \in \mathbb{C}^{L \times L}$ are orthogonal matrices, such that $\mathbf{U}^{\mathrm{H}}\mathbf{U} = \mathbf{V}^{\mathrm{H}}\mathbf{V} = I_{L \times L}$, collecting the left and right singular vectors, and $\mathbf{S} = \mathrm{diag}\{(\sigma_1, \dots \sigma_{\mathrm{L}})\} \in \mathbb{C}^{L \times L}$ is a diagonal matrix collecting the singular values sorted in a decreasing order $\sigma_1 \gg \sigma_2 \gg \cdots \gg \sigma_L$. Now, a compressed approximation of the actual matrix $\widetilde{\mathbf{Y}}$ can be obtained as follows:

$$\widetilde{\mathbf{Y}} \approx \widetilde{\mathbf{U}}\widetilde{\mathbf{S}}\widetilde{\mathbf{V}}^{H}, \tag{4}$$

where $\widetilde{\mathbf{U}} = [\mathbf{u}_1,\ldots,\mathbf{u}_n] \in \mathbb{C}^{K\times\tilde{n}}$ with $\mathbf{u}_i \in \mathbb{C}^{K\times 1}$ and $\widetilde{\mathbf{V}} = [\mathbf{v}_1,\ldots,\mathbf{v}_n] \in \mathbb{C}^{\tilde{n}\times L}$ with $\mathbf{v}_i \in \mathbb{C}^{L\times 1}$ are the reduced left and right-eigenvector matrices collecting only the first $\tilde{n}$ components (i.e., the first $\tilde{n}$ columns of the original matrices $\mathbf{U}$ and $\mathbf{V}$), and $\widetilde{\mathbf{S}} = diag\{(\sigma_1,\ldots,\sigma_{\tilde{n}})\}$ is a reduced diagonal matrix containing the first $\tilde{n}$ singular values.

The above relationship can be used to obtain a compressed representation $\mathbf{Z} \in \mathbb{C}^{\tilde{n}\times L}$ of the original matrix $\mathbf{Y}$, such that:

$$\mathbf{Z} = \widetilde{\mathbf{U}}^{H}\widetilde{\mathbf{Y}} = \widetilde{\mathbf{S}}\widetilde{\mathbf{V}}^{H}, \tag{5}$$

It is important to note the resulting compressed matrix $\mathbf{Z} \in \mathbb{C}^{\tilde{n}\times L}$ is smaller than the original matrix $\widetilde{\mathbf{Y}}$, since usually $\tilde{n} \ll K$. Moreover, the rows of the compressed matrix $\mathbf{Z}$ can be considered to be the realizations of a new set of output variables $\{\mathbf{z}(\mathbf{x}_l)\}_{l=1}^{L}$, with $\mathbf{z}(\mathbf{x}_l) \in \mathbb{C}^{\tilde{n}\times 1}$, which can be considered as the collection of L samples of a compressed $\tilde{n}$-dimensional output variable. Specifically, the element (i, j) of the matrix $\mathbf{Z}$, corresponds to the i-th output of the compressed representation evaluated at the j-th configuration of the input parameters, i.e., $Z_{ij} = z_i(\mathbf{x}_j)$. This means that only $\tilde{n}$ single-output models need to be trained to represent the whole dataset provided by the matrix $\mathbf{Y}$, thus leading to a substantial improvement in the training time. Once a surrogate model for each of the $\tilde{n}$ components of the compressed multi-output representation in $\mathbf{Z}$ is available, the overall compress surrogate can be inexpensively used to predict the system output $\mathbf{y}\left(\overline{\mathbf{x}}\right) \in \mathbb{C}^{K\times 1}$ for a generic test sample $\overline{\mathbf{x}} \in \mathcal{X}$ as follows:

$$y\left(\overline{\mathbf{x}}\right) \approx \mathbf{\mu} + \widetilde{\mathbf{U}}\overline{\mathbf{Z}}, \tag{6}$$

where $\overline{\mathbf{Z}} = \left[z_1\left(\overline{\mathbf{x}}\right),\ldots,z_{\tilde{n}}\left(\overline{\mathbf{x}}\right)\right]^{T} \in \mathbb{C}^{\tilde{n}\times 1}$.

It is important to remark that the compressed representation $\mathbf{z}(\mathbf{x})$ provided by the PCA compression allows approximating the actual data $\mathbf{y}(\mathbf{x})$ with a tunable accuracy depending on the number of PCA components $\tilde{n}$, such that [28]:

$$\left(\frac{\sigma_{\tilde{n}+1}}{\sigma_1}\right)^2 \leq \varepsilon^2, \tag{7}$$

where ε is a given error threshold tuned by the user.

4. Complex Valued Least-Square Support Vector Machine Regression

The LS-SVM regression is a kernel-based regression with an L2 regularizer, which allows constructing a non-parametric model in which the number of unknowns and complexity is independent from the number of input parameters via the solution of a simple linear system, i.e., by solving a convex optimization problem [33].

However, the plain formulation of such a technique has been developed for real-valued datasets only. The aim of this section is to extend the mathematical formulation of the LS-SVM regression to the more general case of the complex-data problem. The proposed formulation is based on the preliminary results presented in [25] and on the results presented in [19,21] for a classical reproducing kernel Hilbert space (RKHS) regression, in which the regularizer term is neglected. For the sake of simplicity, the following complex valued formulation is developed for the simplified case of single-output regression, but all the calculations can be easily extended to the case of a multi-output problem.

Starting from a set of complex-valued samples $\mathcal{D} = \{(\mathbf{x}_l, y_l)\}_{l=1}^{L}$ where $\mathbf{x}_l \in \mathbb{C}^d$ and $y_l = y(\mathbf{x}_l) \in \mathbb{C}$, the primal space formulation of the LS-SVM regression is written:

$$y \approx \widetilde{\mathcal{M}}(\mathbf{x}) = \sum_{i=1}^{N} w_i \phi_i^*(\mathbf{x}) + b = \mathbf{w}, \mathbf{\Phi}(\mathbf{x}) + b, \tag{8}$$

where $\mathbf{w} = [w_1, \ldots, w_N]^T = \mathbf{w}_R + j\mathbf{w}_I \in \mathbb{C}^N$ are the complex regression coefficients (i.e., $w_i = w_{i,R} + jw_{i,I}$), $\mathbf{\Phi}(\mathbf{x}) = \mathbf{\Phi}_R(\mathbf{x}) + j\mathbf{\Phi}_I(\mathbf{x}) = [\phi_1(\mathbf{x}), \ldots, \phi_N(\mathbf{x})]^T$ is a vector-valued complex function $\mathbf{\Phi}(\cdot) : \mathbb{C}^d \to \mathbb{C}^N$ collecting the complex-valued basis functions $\phi_i(\mathbf{x})$ that provide a map between the parameter space and the feature space, $b = b_R + jb_I$ is the bias term, and $\mathbf{w}, \mathbf{\Phi}(\mathbf{x}) = \mathbf{\Phi}^H(\mathbf{x})\mathbf{w} = \mathbf{w}^T\mathbf{\Phi}^*(\mathbf{x})$ is the inner product in the complex domain.

The primal space formulation of the complex-valued regression in Equation (8) can be written in term of its real and imaginary components:

$$\widetilde{\mathcal{M}}(\mathbf{x}) = \widetilde{\mathcal{M}}_R(\mathbf{x}) + j\widetilde{\mathcal{M}}_I(\mathbf{x}) = \left(\mathbf{w}_R\mathbf{\Phi}_R^T(\mathbf{x}) + \mathbf{w}_I\mathbf{\Phi}_I^T(\mathbf{x}) + b_R\right) + j\left(\mathbf{w}_I\mathbf{\Phi}_R^T(\mathbf{x}) - \mathbf{w}_R\,\mathbf{\Phi}_I^T(\mathbf{x}) + b_i\right), \tag{9}$$

In the above primal space formulation, the regression unknowns (i.e., the coefficients in $\mathbf{w}_R$ and $\mathbf{w}_I$, and the bias term b_R and b_I, respectively) are estimated by solving the following convex optimization problem:

$$\min_{\mathbf{w}_R,\mathbf{w}_I,b_R,b_I,\mathbf{e}_R,\mathbf{e}_I} \frac{1}{2}\mathbf{w}_R^T\mathbf{w}_R + \frac{1}{2}\mathbf{w}_I^T\mathbf{w}_I + \frac{\gamma_R}{2}\sum_{l=1}^{L} e_{R,l}^2 + \frac{\gamma_I}{2}\sum_{l=1}^{L} e_{I,l}^2 \tag{10}$$

Such that

$$e_{R,l} = Re\left\{y_l - \widetilde{\mathcal{M}}(\mathbf{x}_l)\right\} = y_{R,l} - (\mathbf{w}_R\mathbf{\Phi}_R(\mathbf{x}_l) + \mathbf{w}_I\mathbf{\Phi}_I(\mathbf{x}_l) + b_R)$$

$$e_{R,l} = Im\left\{y_l - \widetilde{\mathcal{M}}(\mathbf{x}_l)\right\} = y_{I,l} - (\mathbf{w}_I\mathbf{\Phi}_R(\mathbf{x}_l) - \mathbf{w}_R\mathbf{\Phi}_I(\mathbf{x}_l) + b_I)$$

For $l = 1, \ldots, L$, where $\mathbf{w}_R^T\mathbf{w}_R + \mathbf{w}_I^T\mathbf{w}_I = w_2^2$ is the L2 regularizer and the terms $e_{R,l}^2$ and $e_{I,l}^2$ provide the error of a squared loss function.

The Lagrangian for the above constraint optimization problem is written:

$$\begin{aligned}
\mathcal{L}(w_R&, w_I, b_R, b_I, e_R, e_I, \alpha_R, \alpha_I) \\
&= \tfrac{1}{2}w_R^T w_R + \tfrac{1}{2}w_I^T w_I + \tfrac{\gamma_R}{2}\sum_{l=1}^{L} e_{R,l}^2 + \tfrac{\gamma_I}{2}\sum_{l=1}^{L} e_{I,l}^2 \\
&\quad - \sum_{l=1}^{L} \alpha_{R,l}\left\{e_{R,l} - y_{R,l} + (w_R^T\mathbf{\Phi}_R(\mathbf{x}_l) + w_I^T\mathbf{\Phi}_I(\mathbf{x}_l) + b_R)\right\} \\
&\quad - \sum_{l=1}^{L} \alpha_{I,l}\left\{e_{I,l} - y_{I,l} + (w_I^T\mathbf{\Phi}_R(\mathbf{x}_l) - w_R^T\mathbf{\Phi}_I(\mathbf{x}_l) + b_I)\right\},
\end{aligned} \tag{11}$$

where $\alpha_l = \alpha_{R,l} + j\alpha_{I,l}$ are the Lagrangian multipliers such that $\alpha_{R,l}, \alpha_{I,l} \geq 0$ for $l = 1, \ldots, L$.

By computing the partial derivatives of the Lagrangian $\mathcal{L}(w_R, w_I, b_R, b_I, e_R, e_I, \alpha_R, \alpha_I)$ with respect to its parameters:

$$\frac{\partial \mathcal{L}}{\partial \mathbf{w}_R} = 0 \rightarrow \mathbf{w}_R = \sum_{l=1}^{L} [\alpha_{R,l}\mathbf{\Phi}_R(\mathbf{x}_l) - \alpha_{I,l}\mathbf{\Phi}_I(\mathbf{x}_l)], \tag{12}$$

$$\frac{\partial \mathcal{L}}{\partial \mathbf{w}_I} = 0 \rightarrow \mathbf{w}_I = \sum_{l=1}^{L} [\alpha_{R,l}\mathbf{\Phi}_I(\mathbf{x}_l) + \alpha_{I,l}\mathbf{\Phi}_R(\mathbf{x}_l)], \tag{13}$$

$$\frac{\partial \mathcal{L}}{\partial b_R} = 0 \rightarrow \sum_{l=1}^{L} \alpha_{R,l} = 0, \tag{14}$$

$$\frac{\partial \mathcal{L}}{\partial b_I} = 0 \rightarrow \sum_{l=1}^{L} \alpha_{I,l} = 0,$$

(15)

$$\frac{\partial \mathcal{L}}{\partial e_{R,l}} = 0 \rightarrow \gamma_R e_{R,l} = \alpha_{R,l},$$

(16)

$$\frac{\partial \mathcal{L}}{\partial e_{I,l}} = 0 \rightarrow \gamma_I e_{I,l} = \alpha_{I,l},$$

(17)

$$\frac{\partial \mathcal{L}}{\partial \alpha_{R,l}} = 0 \rightarrow e_{R,l} - y_{R,l} + \left(\mathbf{w}_R^T \boldsymbol{\Phi}_R(\mathbf{x}_l) + \mathbf{w}_I^T \boldsymbol{\Phi}_I(\mathbf{x}_l) + b_R \right) = 0,$$

(18)

$$\frac{\partial \mathcal{L}}{\partial \alpha_{I,l}} = 0 \rightarrow e_{I,l} - y_{I,l} + \left(\mathbf{w}_I^T \boldsymbol{\Phi}_R(\mathbf{x}_l) - \mathbf{w}_R^T \boldsymbol{\Phi}_I(\mathbf{x}_l) + b_I \right) = 0,$$

(19)

For $l = 1, \ldots, L$.

By substituting (12)–(17) in (18) and (19):

$$\begin{cases} \frac{\alpha_{R,l}}{\gamma_R} - y_{R,l} + \sum_{i=1}^{L} [\alpha_{R,i} \boldsymbol{\Phi}_R(\mathbf{x}_i) - \alpha_{I,i} \boldsymbol{\Phi}_I(\mathbf{x}_i)]^T \boldsymbol{\Phi}_R(\mathbf{x}_l) + \sum_{i=1}^{L} [\alpha_{R,i} \boldsymbol{\Phi}_I(\mathbf{x}_i) + \alpha_{I,i} \boldsymbol{\Phi}_R(\mathbf{x}_i)]^T \boldsymbol{\Phi}_I(\mathbf{x}_l) + b_R = 0 \\[2mm] \frac{\alpha_{I,l}}{\gamma_I} - y_{I,l} + \sum_{i=1}^{L} [\alpha_{R,i} \boldsymbol{\Phi}_I(\mathbf{x}_i) + \alpha_{I,i} \boldsymbol{\Phi}_R(\mathbf{x}_i)]^T \boldsymbol{\Phi}_R(\mathbf{x}_l) - \sum_{i=1}^{L} [\alpha_{R,i} \boldsymbol{\Phi}_R(\mathbf{x}_i) - \alpha_{I,i} \boldsymbol{\Phi}_I(\mathbf{x}_i)]^T \boldsymbol{\Phi}_I(\mathbf{x}_l) + b_I = 0 \\[2mm] \sum_{l=1}^{L} \alpha_{R,l} = 0 \\[2mm] \sum_{l=1}^{L} \alpha_{I,l} = 0 \end{cases}$$

(20)

For $l = 1, \ldots, L$.

The above system of equations is the dual-form representation of the optimization problem in Equation (10), in which the original regression coefficients collected in the vectors $\mathbf{w}_R$ and $\mathbf{w}_I$ are replaced by the Lagrangian multipliers $\boldsymbol{\alpha}_R$ and $\boldsymbol{\alpha}_I$. It is important to remark that, although the number of unknowns in the primal space (i.e., the dimensionality of $|\mathbf{w}| = N$) is given by the number of basis functions, in the dual formulation the number of unknowns (i.e., the Lagrangian multipliers collected in the vectors $\boldsymbol{\alpha} = \boldsymbol{\alpha}_R + j\boldsymbol{\alpha}_I$) is always equal to the number of the training samples L. This means the resulting model is non-parametric, i.e., a model in which its complexity is independent from the number of both the input parameters and the basis functions.

In order to define the kernel function and the dual formulation of the complex-valued LS-SVM, the first two equations in Equation (20) can be rewritten as follows:

$$\frac{\alpha_{R,l}}{\gamma_R} - y_{R,l} + \sum_{i=1}^{L} \alpha_{R,i} \left[\boldsymbol{\Phi}_R^T(\mathbf{x}_i) \boldsymbol{\Phi}_R(\mathbf{x}_l) + \boldsymbol{\Phi}_I^T(\mathbf{x}_i) \boldsymbol{\Phi}_I(\mathbf{x}_l) \right] + \sum_{i=1}^{L} \alpha_{I,i} \left[\boldsymbol{\Phi}_R^T(\mathbf{x}_i) \boldsymbol{\Phi}_I(\mathbf{x}_l) - \boldsymbol{\Phi}_I^T(\mathbf{x}_i) \boldsymbol{\Phi}_R(\mathbf{x}_l) \right] + b_R = 0,$$

(21)

$$\frac{\alpha_{I,l}}{\gamma_I} - y_{I,l} + \sum_{i=1}^{L} \alpha_{R,i} \left[\boldsymbol{\Phi}_I^T(\mathbf{x}_i) \boldsymbol{\Phi}_R(\mathbf{x}_l) - \boldsymbol{\Phi}_R^T(\mathbf{x}_i) \boldsymbol{\Phi}_I(\mathbf{x}_l) \right] + \sum_{i=1}^{L} \alpha_{I,i} \left[\boldsymbol{\Phi}_R^T(\mathbf{x}_i) \boldsymbol{\Phi}_R(\mathbf{x}_l) + \boldsymbol{\Phi}_I^T(\mathbf{x}_i) \boldsymbol{\Phi}_I(\mathbf{x}_l) \right] + b_I = 0,$$

(22)

For $l = 1, \ldots, L$.

Now, let us define a complex-valued kernel $k_c\left(\mathbf{x}, \mathbf{x}'\right)$:

$$\begin{aligned} k_c\left(\mathbf{x}, \mathbf{x}'\right) &= \boldsymbol{\Phi}(\mathbf{x}), \boldsymbol{\Phi}\left(\mathbf{x}'\right) = \boldsymbol{\Phi}^T(\mathbf{x}) \cdot \boldsymbol{\Phi}^*\left(\mathbf{x}'\right) = \left(\boldsymbol{\Phi}_R(\mathbf{x}) + j\boldsymbol{\Phi}_I(\mathbf{x})\right)^T \cdot \left(\boldsymbol{\Phi}_R\left(\mathbf{x}'\right) - j\boldsymbol{\Phi}_I\left(\mathbf{x}'\right)\right) = \\ &\left[\boldsymbol{\Phi}_R^T(\mathbf{x}) \boldsymbol{\Phi}_R\left(\mathbf{x}'\right) + \boldsymbol{\Phi}_I^T(\mathbf{x}) \boldsymbol{\Phi}_I\left(\mathbf{x}'\right) \right] + j \left[\boldsymbol{\Phi}_I^T(\mathbf{x}) \boldsymbol{\Phi}_R\left(\mathbf{x}'\right) - \boldsymbol{\Phi}_R^T(\mathbf{x}) \boldsymbol{\Phi}_I\left(\mathbf{x}'\right) \right] = k_{\mathbb{R}}\left(\mathbf{x}, \mathbf{x}'\right) + j k_{\mathbb{I}}\left(\mathbf{x}, \mathbf{x}'\right) \end{aligned}$$

(23)

Similar to the real-valued formulation, the kernel function is defined as the inner product in the complex feature space of the basis functions evaluated at $\mathbf{x}$ and $\mathbf{x}'$, where:

$$k_{\mathbb{R}}\left(\mathbf{x}, \mathbf{x}'\right) = \boldsymbol{\Phi}_R^T(\mathbf{x}) \boldsymbol{\Phi}_R\left(\mathbf{x}'\right) + \boldsymbol{\Phi}_I^T(\mathbf{x}) \boldsymbol{\Phi}_I\left(\mathbf{x}'\right) \text{ and } k_{\mathbb{I}}\left(\mathbf{x}, \mathbf{x}'\right) = \boldsymbol{\Phi}_I^T(\mathbf{x}) \boldsymbol{\Phi}_R\left(\mathbf{x}'\right) - \boldsymbol{\Phi}_R^T(\mathbf{x}) \boldsymbol{\Phi}_I\left(\mathbf{x}'\right).$$

Using the above definition, Equations (21) and (22) can be rewritten as follows:

$$\frac{\alpha_{R,l}}{\gamma_R} - y_{R,l} + \sum_{i=1}^{L}[\alpha_{R,i}k_{\mathbb{R}}(\mathbf{x}_i, \mathbf{x}_l) - \alpha_{I,i}k_{\mathbb{I}}(\mathbf{x}_i, \mathbf{x}_l)] + b_R = 0 \tag{24}$$

$$\frac{\alpha_{I,l}}{\gamma_I} - y_{I,l} + \sum_{i=1}^{L}[\alpha_{I,i}k_{\mathbb{R}}(\mathbf{x}_i, \mathbf{x}_l) + \alpha_{R,i}k_{\mathbb{I}}(\mathbf{x}_i, \mathbf{x}_l)] + b_I = 0 \tag{25}$$

For $l = 1, \ldots, L$.

In the above equations, the regression unknowns can be computed by solving the following linear system:

$$\begin{bmatrix} K_{RR} + \frac{I_L}{\gamma_R} & K_{RI} & 1 & 0 \\ K_{IR} & K_{II} + \frac{I_L}{\gamma_I} & 0 & 1 \\ \mathbf{1}^T & \mathbf{0}^T & 0 & 0 \\ \mathbf{0}^T & \mathbf{1}^T & 0 & 0 \end{bmatrix} \begin{bmatrix} \boldsymbol{\alpha}_R \\ \boldsymbol{\alpha}_I \\ b_R \\ b_I \end{bmatrix} = \begin{bmatrix} \mathbf{y}_R \\ \mathbf{y}_I \\ 0 \\ 0 \end{bmatrix} \tag{26}$$

where I_L is the identity matrix of size $L \times L$, $\mathbf{1}^T = [1, \ldots, 1]^T \in \mathbb{R}^{1 \times L}$, $\mathbf{0}^T = [0, \ldots, 0]^T \in \mathbb{R}^{1 \times L}$, $\boldsymbol{\alpha}_R$, $\boldsymbol{\alpha}_I \in \mathbb{R}^{L \times 1}$ are vectors collecting the real and imaginary part of the regression coefficients, $b = b_R + jb_I$ is the bias term and K_{RR}, K_{RI}, K_{IR}, $K_{II} \in \mathbb{R}^{L \times L}$ are kernel matrices defined as:

$$K_{RR}^{(i,j)} = K_{II}^{(i,j)} = k_{\mathbb{R}}(\mathbf{x}_i, \mathbf{x}_j) \tag{27}$$

$$K_{IR}^{(i,j)} = -K_{RI}^{(i,j)} = k_{\mathbb{I}}(\mathbf{x}_i, \mathbf{x}_j) \tag{28}$$

for any $i, j = 1, \ldots, L$. The parameters γ_R and γ_I are the regularizer hyperparameters tuned by the user and provide a trade-off between the model flatness and its accuracy [34].

It is important to remark that the above linear system is square and, therefore, if the determinant of the square matrix is different from 0, it always yields a unique solution, which leads to the following dual space formulation for complex-valued of LS-SVM regression:

$$y(\mathbf{x}) = \sum_{l=1}^{L} \alpha_l k_C(\mathbf{x}_l, \mathbf{x}) + b \tag{29}$$

By substituting Equation (23) in the above equation gives:

$$y(\mathbf{x}) = \sum_{l=1}^{L}[(\alpha_{R,l}k_{\mathbb{R}}(\mathbf{x}_l, \mathbf{x}) - \alpha_{I,l}k_{\mathbb{I}}(\mathbf{x}_l, \mathbf{x}) + b_R) + j(\alpha_{R,l}k_{\mathbb{I}}(\mathbf{x}_l, \mathbf{x}) + \alpha_{I,l}k_{\mathbb{R}}(\mathbf{x}_l, \mathbf{x}) + b_I)] \tag{30}$$

From the above formulation, it is clear that by means of the complex kernel k_C, the complex-valued LS-SVM regression in the dual space is able to account for possible correlation between the real and imaginary part of $y(\mathbf{x})$.

4.1. Complex-Valued Kernel

There are several strategies to construct a complex kernel k_C. Within this paper, we will investigate two of them, the independent kernel, referred to as the complex valued complex function (CVCF) [19], and the pseudo kernel, referred to as the pseudo complex-valued function (PCF) [35,36]. A generic CVCF kernel can be constructed starting from a real-valued kernel $k_{\mathbb{R}}$ as follows:

$$k_C\left(\mathbf{x}, \mathbf{x}'\right) = k_{\mathbb{R}}(\mathbf{x}_R, \mathbf{x}'_R) + k_{\mathbb{R}}(\mathbf{x}_I, \mathbf{x}'_I) + j(k_{\mathbb{R}}(\mathbf{x}_R, \mathbf{x}'_I) + k_{\mathbb{R}}(\mathbf{x}_I, \mathbf{x}'_R)), \tag{31}$$

The above complex kernel is fully compliant with the definition provided in Equation (23). The real kernel $k_{\mathbb{R}}$ can be any real kernel function, e.g., linear kernel, radial basis function

(RBF) kernel and polynomial kernel. Hereafter, in this paper, for the CVCF kernel we will use $k_{\mathbb{R}}$ as the RBF kernel, i.e.,

$$k_{\mathbb{R}}\left(\mathbf{x}, \mathbf{x}'\right) = \exp\left(-\frac{1}{2\sigma^2}\mathbf{x} - \mathbf{x}\prime)^2\right), \tag{32}$$

where σ is the kernel hyperparameter, which in this work will be tuned, along with the regularizer hyperparameters, during the model training by combining cross validation (CV) with a Bayesian optimizer [37,38].

As an alternative, a family of kernels based on the PCF can be suitably generated from the isotropic complex covariance function, such that (additional mathematical details are provided in [19,36]):

$$k_c(\mathbf{x}, \mathbf{x}\prime) = \cos\left(c\mathbf{x} - \mathbf{x}'\right)k_{\mathbb{R}}\left(\mathbf{x}, \mathbf{x}'\right) + j\,\sin(c\mathbf{x} - \mathbf{x}\prime)k_{\mathbb{R}}\left(\mathbf{x}, \mathbf{x}'\right), \tag{33}$$

where $k_{\mathbb{R}}(\mathbf{x}, \mathbf{x}_k)$ can be selected as any kernel function and c is a new hyperparameter. In this specific case, a Rational Quadratic kernel is adopted, such as $k_{\mathbb{R}}(\mathbf{x}, \mathbf{x}\prime) = \sigma^2\left(1 + \frac{\mathbf{x}-\mathbf{x}\prime^2}{2al^2}\right)^{-a}$, which is a Rational Quadratic kernel in which a, l and σ are additional hyperparameters. Moreover, in this case all the hyperparameters are tuned via CV and a Bayesian optimizer [37,38].

4.2. Dual Channel Kernel (DCK) LS-SVM for Complex-Valued Data

The dual channel kernel (DCK) formulation can be seen as a special case of the general mathematical framework presented in the previous section. The underlying idea is to recast the complex variables in terms of their real and imaginary part and to work with a standard real kernel, i.e., $k_c = k_{\mathbb{R}} \colon \mathbb{R}^{d\times d} \to \mathbb{R}$.

Under the above assumption, the regression problem in Equation (26) can be simplified as follows:

$$\begin{bmatrix} K_{RR} + \frac{I_L}{\gamma_R} & 0_L & 1 & 0 \\ 0_L & K_{RR} + \frac{I_L}{\gamma_I} & 0 & 1 \\ 1^T & 0^T & 0 & 0 \\ 0^T & 1^T & 0 & 0 \end{bmatrix} \begin{bmatrix} \alpha_R \\ \alpha_I \\ b_R \\ b_I \end{bmatrix} = \begin{bmatrix} y_R \\ y_I \\ 0 \\ 0 \end{bmatrix} \tag{34}$$

where 0_L is the $L \times L$ null matrix, $K_{RR} \in \mathbb{R}^{L\times L}$ such that $K_{RR}^{(i,j)} = k_{\mathbb{R}}\left(\mathbf{x}_i, \mathbf{x}_j\right)$, whilst the matrices $K_{IR} = -K_{RI} = 0_L$.

It is important to remark that, in the above formulation, there is no coupling between the real and imaginary coefficients α_R and α_I, and bias terms b_R and b_I. Indeed, the solution of the above linear system is equivalent to solving two decoupled ones, accounting for the real and imaginary parts of the regression unknowns independently, such as:

$$\begin{bmatrix} K_{RR} + \frac{I_L}{\gamma_R} & 1 \\ 1^T & 0 \end{bmatrix} \begin{bmatrix} \alpha_R \\ b_R \end{bmatrix} = \begin{bmatrix} y_R \\ 0 \end{bmatrix}, \tag{35}$$

$$\begin{bmatrix} K_{RR} + \frac{I_L}{\gamma_I} & 1 \\ 1^T & 0 \end{bmatrix} \begin{bmatrix} \alpha_I \\ b_I \end{bmatrix} = \begin{bmatrix} y_I \\ 0 \end{bmatrix}, \tag{36}$$

In this work, a standard RBF kernel is considered as the real kernel $k_{\mathbb{R}}\left(\mathbf{x}_i, \mathbf{x}_j\right)$. In the above scenario, the dual space formulation of the LS-SVM is written [27]:

$$y_R(\mathbf{x}) = \sum_{l=1}^{L} \alpha_{R,l} k_{\mathbb{R}}(\mathbf{x}_l, \mathbf{x}) + b_R, \tag{37}$$

$$y_I(\mathbf{x}) = \sum_{l=1}^{L} \alpha_{I,l} k_{\mathbb{R}}(\mathbf{x}_l, \mathbf{x}) + b_I, \tag{38}$$

It is important to note that, in the above formulation, the model for the real and imaginary parts of y are built separately, thus ignoring any possible correlation between them. Furthermore, the above models can be trained using the LS-SVM Lab toolbox for the LS-SVM available in MATLAB [39].

5. Application Examples

This section compares the accuracy and the robustness against noise of the three implementations of the complex-valued LS-SVM regression provided in Section 4 by considering two different application examples. Specifically, the proposed approaches are applied to predict the scattering parameters of a serpentine structure with three parameters and the transfer function of a high-speed link with four parameters.

5.1. Example I

As a first test case, the complex-valued LS-SVM regression applied to calculate the scattering parameters of a serpentine delay line structure is presented. Serpentine lines are widely used in printed circuit board (PCB) design to compensate time delays introduced by the trace routing. However, the frequency-domain behavior of such a structure is heavily affected by its geometrical and electrical parameters and should be carefully assessed during the design phase to avoid signal and power integrity issues and to meet design constraints [40].

The structure of the serpentine delay line considered in this example is shown in Figure 1 (inspired by [40]). The S21 scattering parameters of the above structure are investigated as a function of three parameters (i.e., $\mathbf{x} = [\varepsilon_r, LL, SW]^T$) in a frequency bandwidth from 1 MHz to 1 GHz. Table 1 shows the range of variability serpentine delay line parameters that are simulated to produce the training and test data.

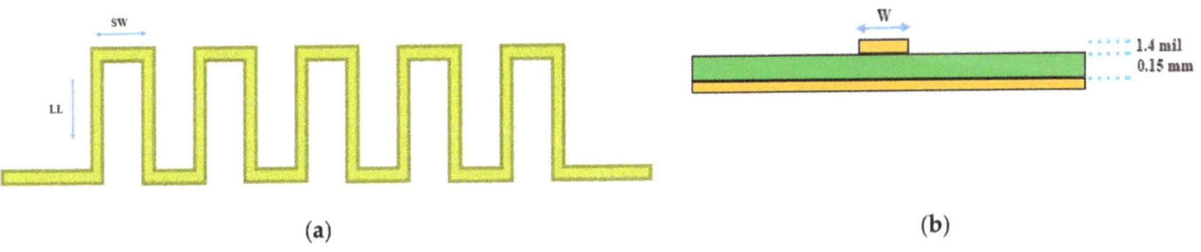

Figure 1. (**a**) Design parameters of the serpentine line to be analyzed; (**b**) cross-sectional view of the serpentine line.

Table 1. Serpentine delay line parameters for the training and test dataset.

Training Ranges	Test Ranges
4.5 mm $\leq LL \leq$ 5.1 mm	4.5 mm $\leq LL \leq$ 5.1 mm
3.9 $\leq \varepsilon_r \leq$ 4.5	3.9 $\leq \varepsilon_r \leq$ 4.5
0.13 mm $\leq SW \leq$ 0.17 mm	0.13 mm $\leq SW \leq$ 0.17 mm
1 MHz $\leq f \leq$ 3 GHz	1 MHz $\leq f \leq$ 3 GHz
1000 samples for each frequency	2000 samples for each frequency

The whole dataset consists of 3000 samples (1000 training data and 2000 test samples). The samples were generated via Latin Hypercube Sampling (LHS) by assuming a uniform variability between their maximum and minimum value. For each configuration of the geometrical parameters, the corresponding scattering parameters were computed for 5000 linearly spaced frequency sample points. The samples were generated via a set of parametric simulations with the full-wave solver available in CST.

The PCA compression is then used to compress the data in the frequency domain and to reduce the computation for the model training. Figure 2 shows the behavior normalized singular values of the frequency points of the training dataset. The plot shows that $\bar{n} = 9$ is enough to represent the whole training set with a 0.001% threshold.

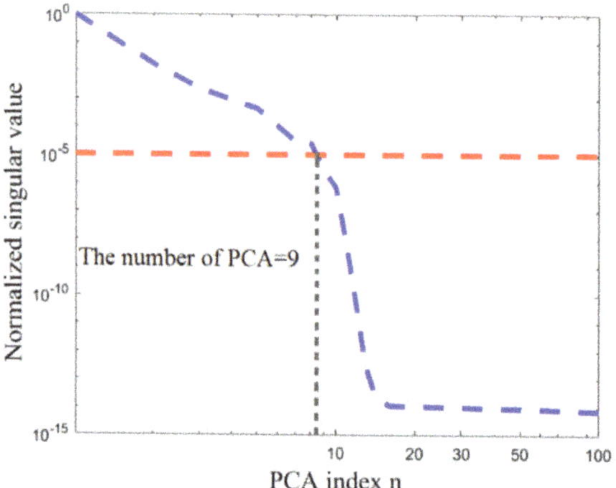

Figure 2. Normalized singular value plot of the serpentine delay line for the considered dataset with 5000 frequency points (blue line). The horizontal line shows the 0.001% threshold for the PCA truncation.

After applying the PCA, the performances of the LS-SVM regression using the PCF and the CVCF complex kernel function (see Section 4.1), and of the DCK LS-SVM regression (see Section 4.2), were assessed on the test samples for the S21 parameter. To investigate the performance of the mentioned methods, the relative Root Mean Square Error (NRMSE) is calculated as the following equation:

$$NRMSE \% = 100 \cdot \frac{\frac{1}{T}\sqrt{\sum_{t=1}^{T}\left(X_y - X_{\hat{y}}\right)^2}}{\frac{1}{T}\sqrt{\sum_{t=1}^{T} X_y^2}}, \tag{39}$$

where X_y can be either the real or imaginary part of the actual test samples, $X_{\hat{y}}$ is the corresponding prediction estimated via the proposed metamodels, and T is the number of test samples.

Figures 3 and 4 show the normalized error for real and imaginary parts of these three regression approaches for an increasing number of training samples (i.e., L = 50, 250, and 500).

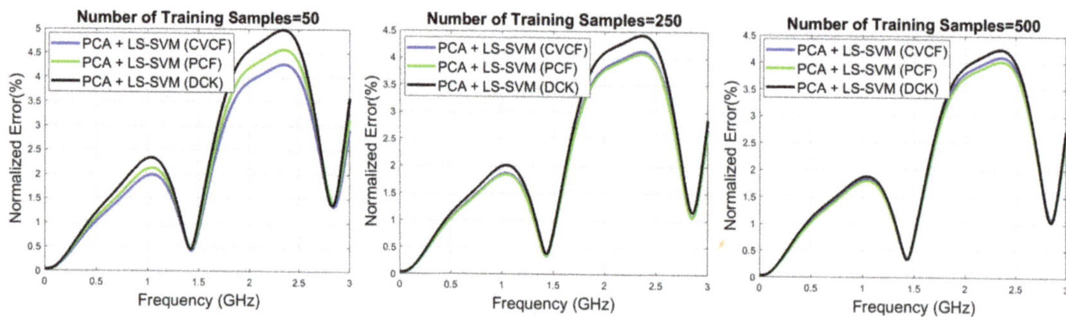

Figure 3. Comparison of the relative NRMSE values computed by the proposed approaches on the test samples by considering the real part of the S21 parameter of the serpentine delay line structure for an increasing number of training samples (i.e., L = 50, 250, 500).

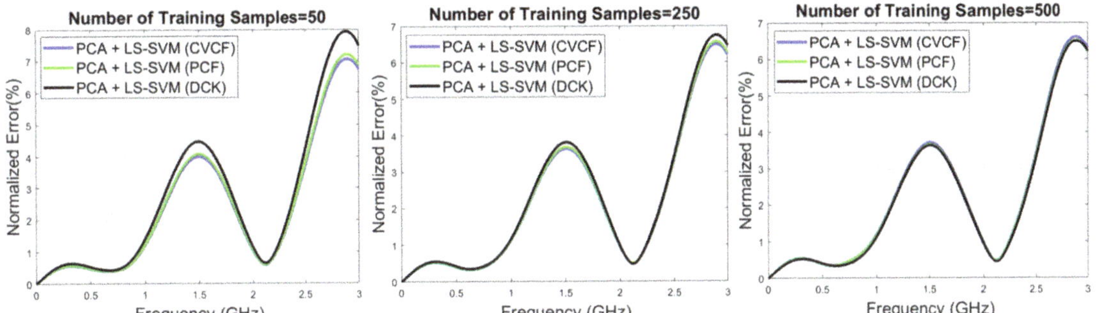

Figure 4. Comparison of the relative NRMSE values computed by the proposed approaches on the test samples by considering the imaginary part of the S21 parameter of the serpentine delay line structure for an increasing number of training samples (i.e., L = 50, 250, 500).

The plots highlight the improved accuracy of the CVCF and PCF with respect to the DCK. Indeed, with the DCK-based model, we are implicitly neglecting any kind of correlation between the real and imaginary parts of the S parameters, and this lack of complexity becomes even more evident when a low number of training samples is used to train the model.

5.2. Example II

The second application example is based on the high-speed link, as depicted in Figure 5, representing a signal distribution on a PCB. Similar to the previous example, the frequency response of the link, and thus its performance, can be greatly influenced by possible variations of its internal parameters [5].

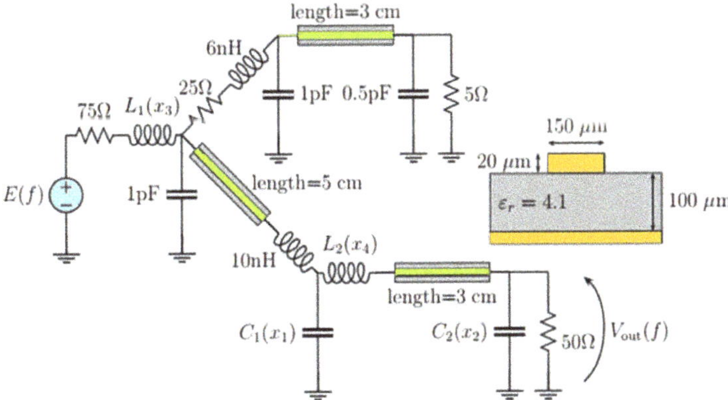

Figure 5. Schematic of the high-speed interconnect link.

Specifically, the proposed modeling approaches are here adopted to build a surrogate model for the frequency-domain behavior of the complex-valued transfer function, in which $y(\mathbf{x}; f) = \frac{Vout(f;\mathbf{x})}{E(f)}$, as a function of the values of four lumped components $C_1(x_1)$, $C_2(x_2)$, $L_1(x_3)$, and $L_2(x_4)$, is defined by four uniformly distributed random variables with a variation of $\pm 50\%$ around their central value collected in $\mathbf{x} = [x_1, x_2, x_3, x_4]^T$ (additional details are provided in Table 2).

Table 2. High-speed interconnect link parameters for the training and test datasets.

Training and Test Ranges	
$C_1\,(x_1)$	$(1 \pm 0.5\,x_1)$ pF
$C_2\,(x_2)$	$(0.5 \pm 0.25\,x_2)$ pF
$L_1\,(x_3)$	$(10 \pm 5\,x_3)$ nH
$L_2\,(x_4)$	$(10 \pm 5\,x_4)$ nH
i	1,2,3,4
x_i	Random variable in $[-1, 1]$

Four sets consisting of L = 20, 100, 150, and 500 training input configurations were generated via an LHS and used as input for a computational model implemented in MATLAB, which provides as output the corresponding transfer function evaluated at 200 frequency points. The PCA is applied to remove redundant information, leading to a compressed representation of the original dataset with only $\bar{n} = 10$ components using a threshold of 0.01%.

The compressed training sets are then used to train three different surrogate models based on the DCK, PCF, and CVCF LS-SVM regressions. For the sake of completeness, the predictions of the above methods are compared with those provided by an additional surrogate model built via a multi-output feedforward NN structure [34,35] considering the real and imaginary parts of the considered transfer function (i.e., using the dual channel implementation). The NN is trained via the Gradient Decent backpropagation method implemented within the Neural Net Fitting Tool available in the MATLAB Deep Learning Toolbox. The network consists of three hidden layers with 50, 20, and 15 neurons, respectively. The activation function is the hyperbolic tangent sigmoid.

Figures 6 and 7 show the performance of each method on a test set consisting of 1000 samples assessed via the relative NRMSE in Equation (39) for the real and imaginary parts, respectively. The results clearly highlight the improved accuracy achieved via the PCF. Moreover, as expected, due to its simplified formulation, the DCK again provides the lowest accuracy. By comparison, NN shows the lower accuracy when a small number of training samples is used (i.e., up to 150 training samples). This low convergence with respect to the number of training samples is due to the inherent non-convex nature of the optimization problem solved during the model training [32].

Moreover, in order to stress the reliability of the proposed techniques, the training output y was corrupted with Gaussian noise, such that:

$$\mathbf{y}_{i,noisy}(\mathbf{x_i}) = \mathbf{y}_i(\mathbf{x_i}) \times (1 + \zeta_n), \tag{40}$$

where $\zeta_n \sim \mathcal{N}\left(0, \sigma_n^2\right)$ is a Gaussian random variable with standard deviation $\sigma_n = [0.01, 0.03]$.

Figures 8 and 9 compare the relative NRMSE computed at a single frequency point selected as the one providing the maximum error via the proposed approaches, and a multi-output feedforward NN for different values of the noise standard deviation σ_n and the number of training samples L for the real and imaginary parts, respectively. Among the introduced methods, CVCF shows the better performance and robustness against noise, both for real and imaginary parts.

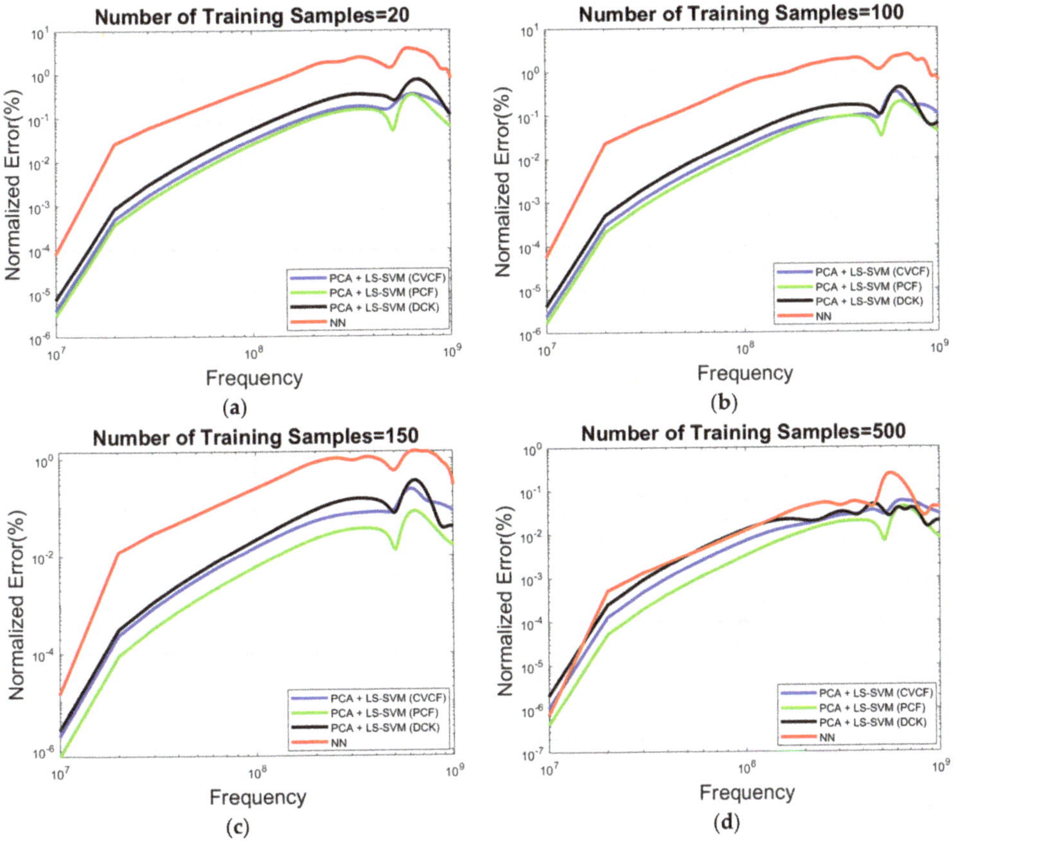

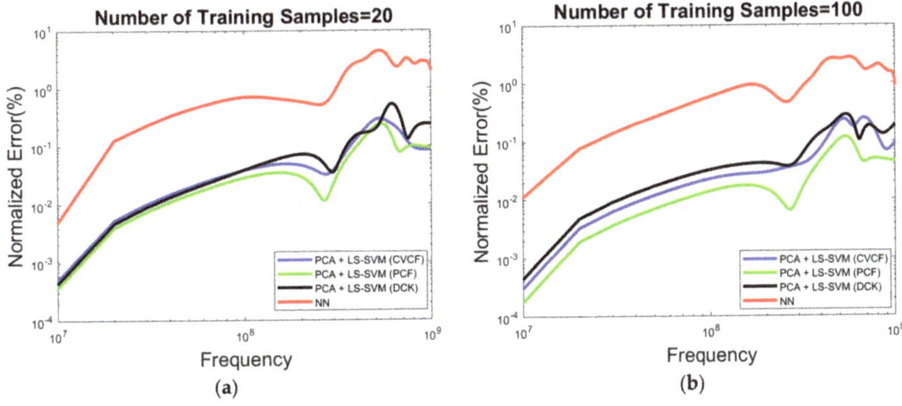

Figure 6. Comparison of the relative NRMSE values computed by the proposed approaches and a feedforward multi-output neural network on the test samples by considering the real part of the transfer function of a high-speed link for an increasing number of training samples (**a**) L = 20; (**b**) L = 100; (**c**) L = 150; (**d**) L = 500.

Figure 7. *Cont.*

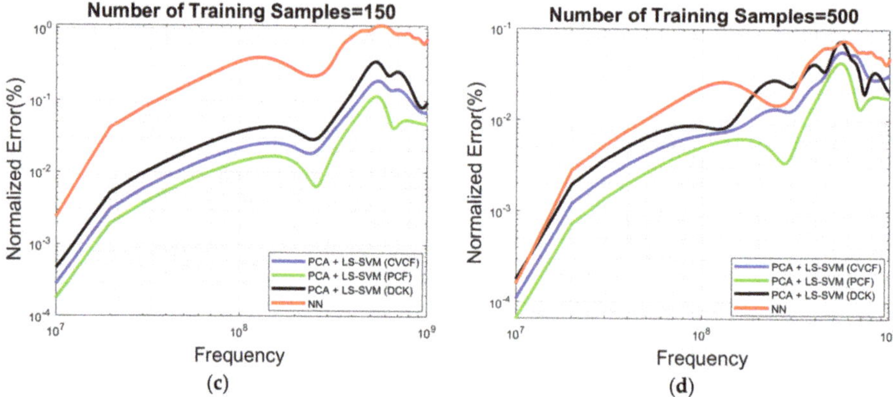

Figure 7. Comparison of the relative NRMSE values computed via the proposed approaches and a feedforward multi-output neural network on the test samples by considering the imaginary part of the transfer function of high-speed link for an increasing number of training samples. (**a**) L = 20; (**b**) L = 100; (**c**) L = 150; (**d**) L = 500).

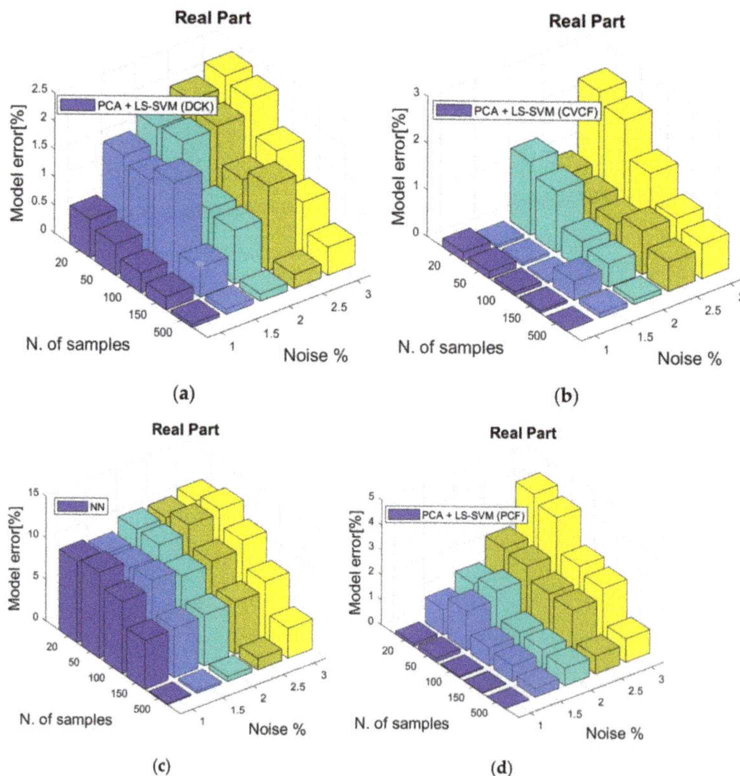

Figure 8. Three-dimensional plot of the relative NRMSE computed on the real part of the test samples at a single frequency point selected as the one providing the maximum error via the proposed approaches (see panel (**a**) for the DCK, panel (**b**) for CVCF and panel (**d**) for PCF), and a multi-output feedforward NN (see panel (**c**))for different values of the noise standard deviation σ_n and the number of training samples L.

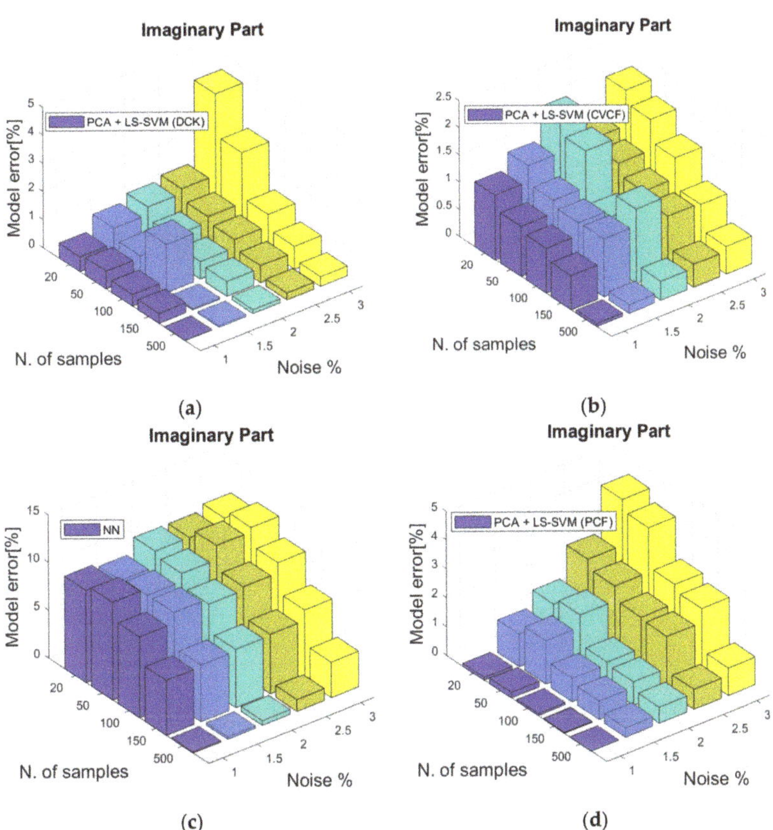

Figure 9. Three-dimensional plot of the relative NRMSE computed on the real part of the test samples at a single frequency point selected as the one providing the maximum error via the proposed approaches (see panel (**a**) for the DCK, panel (**b**) for CVCF and panel (**d**) for PCF), and a multi-output feedforward NN (see panel (**c**)) for different values of the noise standard deviation σ_n and the number of training samples L.

6. Conclusions

In this paper, the application of different methods based on LS-SVM regression for predicting the complex-valued frequency response was presented in a serpentine delay line and a high-speed link. The required data for training and examination were obtained from CST and MATLAB software, in terms of physical and geometrical parameters, including substrate dielectric constant and height and strip width of the serpentine delay line, and the capacitances and inductances of the high-speed link. PCA is shown to reduce the computational cost of the frequency samples. Then, the training data are used to create the model, and the test data are considered for the examination of the established model. The performance of the three proposed methods and of a multi-output feedforward NN are compared on the test dataset. It is shown that the CVCF method is suitable for the first complex electromagnetic problem when there are few training data points. In addition, PCF shows an acceptable performance under noise-free conditions at all frequency points for the second example. Finally, the results show that CVCF is the best method when the level of noise is increased.

Author Contributions: Conceptualization, N.S. and R.T.; methodology, N.S. and R.T.; software, data curation and validation, N.S.; writing original draft preparation, N.S.; writing review and editing, N.S. and R.T.; supervision, R.T. All authors have read and agreed to the published version of the manuscript.

Funding: This research received no external funding.

Acknowledgments: The authors would like to thank Flavio Canavero, Politecnico di Torino, Italy, for his valuable and constructive suggestions during the planning and development of this manuscript.

Conflicts of Interest: The authors declare no conflict of interest.

References

1. Manfredi, P.; Ginste, D.V.; Stievano, I.S.; de Zutter, D.; Canavero, F.G. Stochastic transmission line analysis via polynomial chaos methods: An overview. *IEEE Electromagn. Compat. Mag.* **2017**, *6*, 77–84. [CrossRef]
2. Zhang, Z.; El-Moselhy, T.A.; Elfadel, I.M.; Daniel, L. Stochastic testing method for transistor-level uncertainty quantification based on generalized polynomial chaos. *IEEE Trans. Comput. -Aided Des. Integr. Circuits Syst.* **2013**, *32*, 1533–1545. [CrossRef]
3. Spina, D.; Ferranti, F.; Dhaene, T.; Knockaert, L.; Antonini, G.; Ginste, D.V. Variability analysis of multiport systems via polynomial-chaos expansion. *IEEE Trans. Microw. Theory Tech.* **2012**, *60*, 2329–2338. [CrossRef]
4. Ahadi, M.; Roy, S. Sparse Linear Regression (SPLINER) Approach for Efficient Multidimensional Uncertainty Quantification of High-Speed Circuits. *IEEE Trans. Comput. -Aided Des. Integr. Circuits Syst.* **2016**, *35*, 1640–1652. [CrossRef]
5. Trinchero, R.; Manfredi, P.; Stievano, I.S.; Canavero, F.G. Machine Learning for the Performance Assessment of High-Speed Links. *IEEE Trans. Electromagn. Compat.* **2018**, *60*, 1627–1634. [CrossRef]
6. Ma, H.; Li, E.; Cangellaris, A.C.; Chen, X. Support Vector Regression-Based Active Subspace (SVR-AS) Modeling of High-Speed Links for Fast and Accurate Sensitivity Analysis. *IEEE Access* **2020**, *8*, 74339–74348. [CrossRef]
7. Treviso, F.; Trinchero, R.; Canavero, F.G. Multiple delay identification in long interconnects via LS-SVM regression. *IEEE Access* **2021**, *9*, 39028–39042. [CrossRef]
8. Houret, T.; Besnier, P.; Vauchamp, S.; Pouliguen, P. Controlled Stratification Based on Kriging Surrogate Model: An Algorithm for Determining Extreme Quantiles in Electromagnetic Compatibility Risk Analysis. *IEEE Access* **2020**, *8*, 3837–3847. [CrossRef]
9. Watson, P.M.; Gupta, K.C.; Mahajan, R.L. Development of knowledge based artificial neural network models for microwave components. *IEEE Int. Microw. Symp. Baltim.* **1998**, *1*, 9–12.
10. Veluswami, A.; Nakhla, M.S.; Zhang, Q.-J. The application of neural networks to EM based simualtion and optimization of interconencts in high-speed VLSI circuits. *IEEE Trans. Microw. Theory Techn.* **1997**, *45*, 712–723. [CrossRef]
11. Kumar, R.; Narayan, S.L.; Kumar, S.; Roy, S.; Kaushik, B.K.; Achar, R.; Sharma, R. Knowledge-Based Neural Networks for Fast Design Space Exploration of Hybrid Copper-Graphene On-Chip Interconnect Networks. *IEEE Trans. Electromagn. Compat.* **2021**, 1–14, *Early Access Article*. [CrossRef]
12. Swaminathan, M.; Torun, H.M.; Yu, H.; Hejase, J.A.; Becker, W.D. Demystifying Machine Learning for Signal and Power Integrity Problems in Packaging. *IEEE Trans. Compon. Packag. Manuf. Technol.* **2020**, *10*, 1276–1295. [CrossRef]
13. Jin, J.; Zhang, C.; Feng, F.; Na, W.; Ma, J.; Zhang, Q. Deep Neural Network Technique for High-Dimensional Microwave Modeling and Applications to Parameter Extraction of Microwave Filters. *IEEE Trans. Microw. Theory Tech.* **2019**, *67*, 4140–4155. [CrossRef]
14. Moradi, M.; Sadrossadat, A.; Derhami, V. Long Short-Term Memory Neural Networks for Modeling Nonlinear Electronic Components. *IEEE Trans. Compon. Packag. Manuf. Technol.* **2021**, *1*, 840–847. [CrossRef]
15. Bourinet, J.-M. *Reliability Analysis and Optimal Design under Uncertainty—Focus on Adaptive Surrogate-Based Approaches*; Computation [stat.CO]; Université Clermont Auvergne: Clement Ferrand, France, 2018.
16. Scardapane, S.; van Vaerenbergh, S.; Hussain, A.; Uncini, A. Complex-valued neural networks with nonparametric activation functions. *IEEE Trans. Emerg. Top. Comput. Intell.* **2018**, *4*, 140–150. [CrossRef]
17. Adali, T.; Schreier, P.J.; Scharf, L.L. Complex-valued signal processing: The proper way to deal with impropriety. *IEEE Trans. Signal Processing* **2011**, *59*, 5101–5125. [CrossRef]
18. Papaioannou, A.; Zafeiriou, S. Principal Component Analysis with Complex Kernels, IEEE Transactions on Neural Networks and Learning Systems, 2013. pp. 1719–1726. Available online: http://citeseerx.ist.psu.edu/viewdoc/download?doi=10.1.1.402.3888&rep=rep1&type=pdf (accessed on 28 December 2021).
19. Boloix-Tortosa, R.; Murillo-Fuentes, J.J.; Santos, I.; Pérez-Cruz, F. Widely linear complex-valued kernel methods for regression. *IEEE Trans. Signal Processing* **2017**, *65*, 5240–5248. [CrossRef]
20. Tobar, F.A.; Kuh, A.; Mandic, D.P. A novel augmented complex valued kernel LMS. In Proceedings of the 2012 IEEE 7th Sensor Array and Multichannel Signal Processing Workshop (SAM), Hoboken, NJ, USA, 17–20 June 2012; pp. 473–476.
21. Boloix-Tortosa, R.; Murillo-Fuentes, J.J.; Tsaftaris, S.A. The Generalized Complex Kernel Least-Mean-Square Algorithm. *IEEE Trans. Signal Processing* **2019**, *67*, 5213–5222. [CrossRef]
22. Hirose, A. *Complex-Valued Neural Networks*; Springer Science & Business Media: Berlin/Heidelberg, Germany, 2013.
23. Bouboulis, P.; Theodoridis, S.; Mavroforakis, C.; Evaggelatou-Dalla, L. Complex Support Vector Machines for Regression and Quaternary Classification. *IEEE Trans. Neural Netw. Learn. Syst.* **2015**, *26*, 1260–1274. [CrossRef]

24. Scardapane, S.; van Vaerenbergh, S.; Comminiello, D.; Uncini, A. Widely Linear Kernels for Complex-Valued Kernel Activation Functions. In Proceedings of the ICASSP 2019-2019 IEEE International Conference on Acoustics, Speech and Signal Processing (ICASSP), Brighton, UK, 12–17 May 2019; pp. 8528–8532.

25. Ogunfunmi, T.; Paul, T.K. On the complex kernel-based adaptive filter. In Proceedings of the 2011 IEEE International Symposium of Circuits and Systems (ISCAS), Rio de Janeiro, Brazil, 15–18 May 2011; pp. 1263–1266.

26. Boloix-Tortosa, R.; Murillo-Fuentes, J.J.; Payán-Somet, F.J.; Pérez-Cruz, F. Complex Gaussian processes for regression. *IEEE Trans. Neural Netw. Learn. Syst.* **2018**, *29*, 5499–5511. [CrossRef]

27. Soleimani, N.; Trinchero, R.; Canavero, F. Application of Different Learning Methods for the Modelling of Microstrip Characteristics. In Proceedings of the 2020 IEEE Electrical Design of Advanced Packaging and Systems (EDAPS), Shenzhen, China, 14–16 December 2020; pp. 1–3.

28. Manfredi, P.; Grivet-Talocia, S. Compressed Stochastic Macromodeling of Electrical Systems via Rational Polynomial Chaos and Principal Component Analysis. In Proceedings of the 2021 Asia-Pacific International Symposium on Electromagnetic Compatibility (APEMC), Nusa Dua-Bali, Indonesia, 27–30 September 2021.

29. Ahmadi, M.; Sharifi, A.; Fard, M.J.; Soleimani, N. Detection of brain lesion location in MRI images using 326 convolutional neural network and robust PCA. *Int. J. Neurosci.* **2021**, *131*, 1–12.

30. Jolliffe, I.T. *Principal Component Analysis*; Springer: New York, NY, USA, 2002.

31. Manfredi, P.; Trinchero, R. A data compression strategy for the efficient uncertainty quantification of time-domain circuit responses. *IEEE Access* **2020**, *8*, 92019–92027. [CrossRef]

32. Kushwaha, S.; Attar, A.; Trinchero, R.; Canavero, F.; Sharma, R.; Roy, S. Fast Extraction of Per-Unit-Length Parameters of Hybrid Copper-Graphene Interconnects via Generalized Knowledge Based Machine Learning. In Proceedings of the IEEE 30th Conference on Electrical Performance of Electronic Packaging and Systems (EPEPS), Austin, TX, USA, 17–20 October 2021.

33. Suykens, J.A.K.; van Gestel, T.; de Brabanter, J.; de Moor, B.; Vandewalle, J. *Least Squares Support Vector Machines*; World Scientific Publishing Company: Singapore, 2002.

34. Vapnik, V. *The Nature of Statistical Learning Theory*, 2nd ed.; Springer: New York, NY, USA, 1999.

35. Posa, D. Parametric families for complex valued covariance functions: Some results, an overview and critical aspects. *Spat. Stat.* **2020**, *39*, 100473. [CrossRef]

36. Iaco, D.S.; Palma, M.; Posa, D. Covariance functions and models for complex-valued random fields. *Stoch. Environ. Res. Risk Assess.* **2003**, *17*, 145–156. [CrossRef]

37. Snoek, J.; Larochelle, H.; Adams, R.P. Practical bayesian optimization of machine learning algorithms. *Adv. Neural Inf. Processing Syst.* **2012**, *25*, 1–10.

38. Geng, J.; Gan, W.; Xu, J.; Yang, R.; Wang, S. Support vector machine regression (SVR)-based nonlinear modeling of 354 radiometric transforming relation for the coarse-resolution data-referenced relative radiometric normalization (RRN). *Geo-Spat. Inf. Sci.* **2020**, *23*, 237–247. [CrossRef]

39. *LS-SVMlab*, Version 1.8 ed; Department of Electrical Engineering (ESAT), Katholieke Universiteit Leuven: Leuven, Belgium, 6 July 2011; Available online: https://www.esat.kuleuven.be/sista/lssvmlab/old/toolbox.html (accessed on 28 December 2021).

40. Soh, W.-S.; See, K.-Y.; Chang, W.-Y.; Oswal, M.; Wang, L.-B. Comprehensive analysis of serpentine line design. In Proceedings of the 2009 Asia Pacific Microwave Conference, Singapore, 7–10 December 2009; pp. 1285–1288.

Article

Finite-Size and Illumination Conditions Effects in All-Dielectric Metasurfaces

Luca Ciarella [1,†], Andrea Tognazzi [2,3,*,†], Fabio Mangini [4,†], Costantino De Angelis [3,4,5], Lorenzo Pattelli [6,7,†] and Fabrizio Frezza [1,5]

[1]   Dipartimento di Ingegneria dell'Informazione, Elettronica e Telecomunicazioni, Sapienza Università di Roma, Via Eudossiana 18, 00184 Roma, Italy; ciarella.1656100@studenti.uniroma1.it (L.C.); fabrizio.frezza@uniroma1.it (F.F.)

[2]   Dipartimento di Ingegneria, Università degli Studi di Palermo, Viale delle Scienze ed. 8, 90128 Palermo, Italy

[3]   Istituto Nazionale di Ottica-Consiglio Nazionale delle Ricerche (INO-CNR), Via Branze 45, 25123 Brescia, Italy; costantino.deangelis@unibs.it

[4]   Dipartimento di Ingegneria dell'Informazione, Università degli Studi di Brescia,Via Branze 38, 25123 Brescia, Italy; fabio.mangini@unibs.it

[5]   Consorzio Nazionale Interuniversitario per le Telecomunicazioni (CNIT), Viale G.P. Usberti 181/A Sede Scientifica di Ingegneria-Palazzina 3, 43124 Parma, Italy

[6]   Laboratorio Europeo di Spettroscopia Nonlineare (LENS), Via Nello Carrara 1, 50019 Sesto Fiorentino, Italy; pattelli@lens.unifi.it

[7]   Istituto Nazionale di Ricerca Metrologica (INRiM), Strada delle Cacce 91, 10135 Torino, Italy

*   Correspondence: andrea.tognazzi@unipa.it

†   These authors contributed equally to this work.

Abstract: Dielectric metasurfaces have emerged as a promising alternative to their plasmonic counterparts due to lower ohmic losses, which hinder sensing applications and nonlinear frequency conversion, and their larger flexibility to shape the emission pattern in the visible regime. To date, the computational cost of full-wave numerical simulations has forced the exploitation of the Floquet theorem, which implies infinitely periodic structures, in designing such devices. In this work, we show the potential pitfalls of this approach when considering finite-size metasurfaces and beam-like illumination conditions, in contrast to the typical infinite plane-wave illumination compatible with the Floquet theorem.

Keywords: all-dielectric metasurfaces; multipolar decomposition; T-matrix; BIC

Citation: Ciarella, L.; Tognazzi, A.; Mangini, F.; De Angelis, C.; Pattelli, L.; Frezza, F. Finite-Size and Illumination Conditions Effects in All-Dielectric Metasurfaces. *Electronics* **2022**, *11*, 1017. https://doi.org/10.3390/electronics11071017

Academic Editor: Longqing Cong

Received: 14 February 2022
Accepted: 21 March 2022
Published: 24 March 2022

Publisher's Note: MDPI stays neutral with regard to jurisdictional claims in published maps and institutional affiliations.

1. Introduction

All-dielectric metasurfaces have recently emerged as a valid and promising alternative to their plasmonic counterparts. Their lower dissipative losses and thermal heating at optical frequencies allow working at higher pump power for nonlinear frequency generation and the performance of non-invasive sensing of biological samples [1–12]. Moreover, in all-dielectric resonators, electric and magnetic resonances can be used to engineer near and far fields, thus providing more degrees of freedom in the design of optical devices [1,13–16]. In this framework, III-V compounds have been used for second-harmonic generation, due to their non-centrosymmetric structure and their high second-order nonlinear coefficients [2–5,17–23]. Strong third-harmonic generation has been demonstrated instead with Si-based platforms [24,25]. Lately, great efforts have been made to improve the design of optical nanoresonators, either in the limit of single isolated nanoantennas or in the opposite scenario where perfect periodicity is assumed and thus the finite-size real-world structure is approximated by an infinite structure [26–28]. The two opposite situations are required in order to limit the numerical burden in the modelling when resorting to the full-wave solution of Maxwell's equations. Despite the fact that this has given very helpful guidelines, these approaches fail in the accurate design of a device, because they unavoidably neglect the effects that are due to the size of a real device, which is necessarily finite and is often pumped by input beams exciting a limited portion of the device (see Figure 1). Such limitations become particularly relevant when considering metasurfaces supporting bound states in the continuum (BIC), which are non-radiating resonant modes in an open system that are not coupled to the radiating channels and exist only in infinite structures [29]. A common approach to obtain extremely high Q-factor resonances is to introduce a defect into a BIC-supporting metasurface (e.g., by breaking its symmetry), thus realizing a so-called quasi-BIC. This approach is mainly dictated by the fact that currently available simulation tools allow only infinite structures to be inspected, and thus it is necessary to break the symmetry in order to couple to such modes. Quasi-BICs, however, are expected to arise naturally by simply considering the finite extent of a real metasurface.

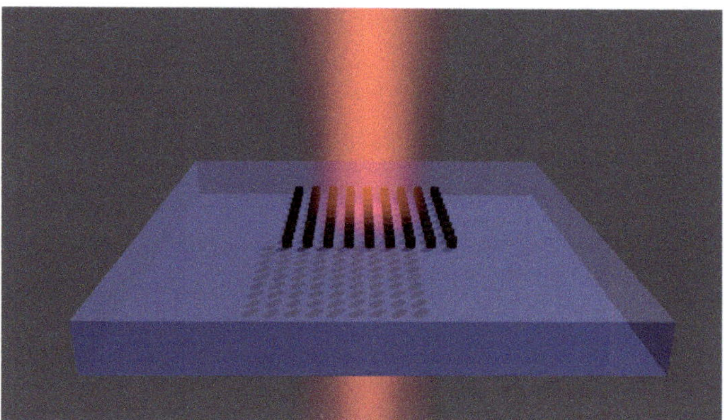

Figure 1. Sketch of a finite-size metasurface on a substrate illuminated by a Gaussian beam. Only a portion of the array is covered by the impinging light.

The necessity of using a finite-size model arises from the fact that the currently adopted models simulate an elementary unit cell and exploit the periodicity of the structure through Floquet's theorem to obtain the optical parameters of an infinite extended metasurface. The results of these models are very good, but they do not take into account the finiteness of the structure and the border effects. Furthermore, when Floquet's theorem is employed, the only possible excitation is a plane wave, unlike in experiments, where a Gaussian beam is usually employed. The combination of the two types of models can help to better foresee the behaviour of an actual metasurface in the experimental phase.

The differences between an infinite model and a finite one are highlighted in [30] and [31], where the situation is analysed with a dipole coupling model for a 2D array of nanoparticles. However, dielectric resonators often support higher-order resonances which cannot be described in the dipole approximation. In order to overcome this limitation, it is necessary to account for higher-order multipolar components, which are of particular interest for dielectric metasurfaces and the study of metamaterials in general [26,32–35]. Babicheva and Evlyukhin [36] provided an analytical model to include a quadrupolar contribution, but the authors employed an infinite structure approximation. In [37], the authors employed the superposition T-matrix scheme with the Ewald sum formalism to describe periodic structures with large unit cells. In [38], the authors studied finite-size effects in plasmonic metasurfaces, highlighting the full-width half-maximum dependence of resonances upon the number of resonators.

Our study aims to highlight the importance of the spatial finiteness of a metasurface supporting high-order multipole components. For this purpose, we employed SMUTHI [39], which implements a linear space-limited model, and tested its performance considering a metasurface composed of silicon nanopillars on a silicon oxide substrate. We analysed the total reflectivity behaviour as a function of the incident wavelength and compared it with the results obtained with the COMSOL infinite periodic model. The reflectivity of the metasurface strongly depends on the number of elements and converges to the infinite array approximation for a large array size, thus highlighting the importance of considering the finiteness of the metasurface. Furthermore, this approach allows the metasurface to be illuminated with a Gaussian beam of a different size under the paraxial approximation, which is not possible when employing Floquet's theorem.

2. Methods

We simulated an all-dielectric metasurface composed of Si cylinders on a SiO_2 substrate, both with a periodic model using COMSOL (exploiting the periodicity via Floquet's theorem) and a space-limited model using SMUTHI, which allowed us to simulate very large finite structures due to the lower computational power required. To test the finite model, we calculated the reflectivity (R) for both models, varying the wavelength of the incident wave λ from 1250 nm to 1800 nm, in steps of 10 nm.

The finite structure was composed of a maximum of 169 cylinders lying on the substrate in a 13×13 configuration, as seen in Figure 2. The refractive indices were 1 for the air, 3.45 for the cylinders, and 1.45 for the substrate, the last two values being the silicon and silicon oxide indices at the working frequencies. The cylinder heights were 428 nm and their radii were 214 nm. The distance between two adjacent cylinders was 856 nm. These dimensions were chosen to create a structure that supports resonant modes at the working frequencies on two levels: inside single cylinders and globally due to the periodicity. The layers consisted of a substrate with thickness $2\lambda / n_{\text{substrate}}$ and an air layer 2λ thick, and the domain on the plane of the cylinders was a square with a side length of 15,500 nm. Although SMUTHI allows the simulation of lossy media, in this study we employed a fixed real value for the refractive indices, because in the spectral region of interest the silicon and silicon dioxide absorption coefficients and the dispersion were very small and thus could be neglected.

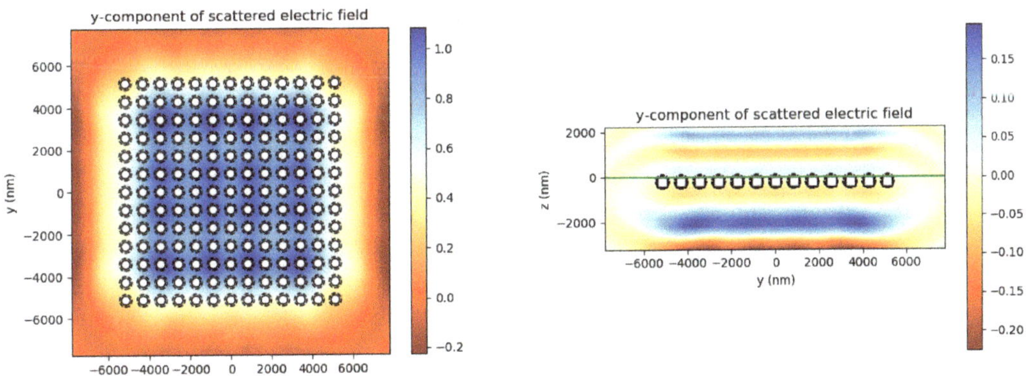

Figure 2. The y-component of the scattered electric field given by SMUTHI in the xy (**left**) and yz planes (**right**) for the structure composed of 13×13 cylinders. The substrate (positive z-axis) has $n = 1.45$ and the air layer (negative z-axis) has $n = 1$. The cylinders have $n = 3.45$, height 428 nm, and radius 214 nm, and 856 nm is the periodicity (distance between two cylinders on the same column or row). The incident wave is a Gaussian beam 15 μm wide at 1600 nm, linearly polarized along the y-axis with an initial maximum intensity of the electric field of $1\,\mathrm{V\,m^{-1}}$, with the maximum at the origin $(x, y, z) = (0, 0, 0)$. The incidence is perpendicular to the substrate and the propagation direction is towards the positive z-direction. The domain on the xy plane is a square with a side length of 15,500 nm.

The incident wave used for the COMSOL periodic model is always a plane wave polarized along the y-axis, while with SMUTHI it is also possible to simulate an incident

Gaussian beam. We set an incident wave propagating from the air layer towards the substrate at normal incidence, at a wavelength which varied as previously described. The polarization of the wave was linear, with the electric field oscillating along the y-axis and an initial intensity of $E_0 = 1\,\mathrm{V\,m^{-1}}$. The maximum value of the electric field was at the air–substrate interface. To calculate the reflectivity, we wrote a MATLAB script, which takes as an input the SMUTHI results related to the scattered field from the cylinders and the reflected one from the substrate. Then, we calculated the flux of the Poynting vector through three surfaces xy, yz, and xz at the border of the domain described above, inside the air layer.

The formulas used to evaluate the Poynting vector components were the following [40]:

$$p_x = \frac{|E_y|^2 + |E_z|^2}{2Z} \qquad\qquad p_y = \frac{|E_x|^2 + |E_z|^2}{2Z}$$

$$p_z = \frac{|E_x|^2 + |E_y|^2}{2Z} \qquad\qquad p_0 = \frac{|E_{0y}|^2}{2Z} = \frac{|E_0|^2}{2Z}$$

where E_x, E_y, and E_z are the field components given by the combinations of the scattered and the reflected waves, E_0 is the initial intensity, and Z is the characteristic impedance of vacuum. Integrating these values along the respective surfaces, it is possible to calculate the reflectivity:

$$R = \frac{P_z + 2P_x + 2P_y}{P_0}$$

where capital P_x, P_y, and P_z are the powers relative to the power densities with the same subscript and are obtained through surface integration over the simulation domain walls. This procedure was repeated for each wavelength to obtain the full spectrum.

3. Validation

To develop the space-limited model we used SMUTHI, a recently developed open-source Python package based on numerical methods, to speed up the computation of light scattering problems [41,42] (an analytical treatment is reported in Appendix A).

To determine how the software performs in our situation, we developed two simple models of the metasurface, one using COMSOL and the other using SMUTHI, to compare the outputs for a structure composed of nine cylinders arranged in a 3×3 configuration around the origin. The parameters were the same as those set in the previous section, except for the domain width in the plane of the cylinders, which was changed to $5000\,\mathrm{nm}$, and the incident plane wave that was fixed at $1200\,\mathrm{nm}$. In COMSOL, we simulated a larger domain: a cube with a side length of $6500\,\mathrm{nm}$ with a perfectly matched layer all around, in order to prevent unwanted reflections from the side. We performed the integration on the same surface as that employed in SMUTHI. We employed the iterative GMRES solver in COMSOL and set a mesh size of $\lambda_e/7$, where λ_e is the effective refractive index of the material.

Figure 3 shows the electric fields at a distance of $1430\,\mathrm{nm}$ from the last cylinder, obtained in COMSOL (left) and SMUTHI (right), respectively.

A good agreement was found using a multipolar expansion with $l = m = 3$ for the cylinders, which resulted in very similar field distributions and integrated reflectance values of:

$$R_{\mathrm{COMSOL}} = 0.0649$$
$$R_{\mathrm{SMUTHI}} = 0.0542$$

Then, for the transmission coefficient:

$$T_{\text{SMUTHI}} = 0.9482$$

and thus the condition $R + T = 1$ is almost respected (this must hold because the refractive indices are real) [41,42].

We underline the times required for the two simulations: for the structure reported in this chapter, COMSOL required 16 min with a reduced geometry, obtained by exploiting symmetry considerations, and 74 min with the full structure, to solve the problem; on the other hand, SMUTHI required 3 min (PC specifications: CPU Intel© Xeon© E5-2650 0, 2 GHz, 8 core. RAM DIMM-DD3 1600 MHz, 64 GB. GPU NVIDIA Quadro K2000). Although COMSOL provides a more complete simulation by computing the field in the full volume of the structure, we only need the field on three specific surfaces, and thus SMUTHI is much more time-efficient.

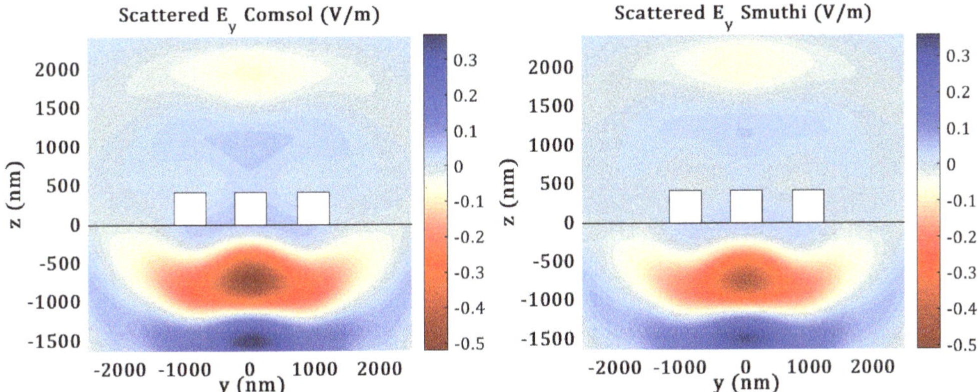

Figure 3. The y-component of the scattered electric field obtained with COMSOL (**left**) and with SMUTHI (**right**) for the structure composed of 3×3 cylinders. The cylinders have $n = 3.45$, height 428 nm, and radius 214 nm, with 856 nm as the periodicity. The incident wave is a plane wave at 1200 nm, linearly polarized along the y-axis at normal incidence. The domain on the xy plane is a square with a side length of 5000 nm. The plots are along the yz plane at a distance of 1430 nm from the last cylinders (2500 nm from the centre of the structure).

4. Results

Figure 4a shows the plots of R, calculated on the same surface, as a function of the wavelength, for different metasurface sizes. When the number of cylinders increases, the spectral features related to the multipolar resonances inside the cylinders and their interplay become more evident. We note that for smaller arrays, the covered area inside the integration surface is much smaller, thus resulting in smaller reflectivity values.

In Figure 4b–d we compare the reflectivity as a function of the wavelength obtained with the two models.

For the finite-size model, the values are relative to the 13×13 cylinder structure and are normalized to have a maximum equal to one. The normalization is only for comparison purposes. Indeed, in the finite structure case, when the integration domain is enlarged, the unpatterned surface increases, thus reducing the reflectivity. However, the relative size of the peaks does not change as this is only due to the cylinders. The normalization of the finite structure to the infinite structure allows the relative intensities of the peaks to be compared, and thus the finite-size effect to be evaluated. We compared the spectra with an incident plane wave for both structures and then changed the wave for the finite model into a Gaussian beam 15 µm and 6 µm wide. Our simulations showed that as long as the paraxial simulation holds and the beam covers the metasurface, the plane wave approximation deviates slightly from the real situation.

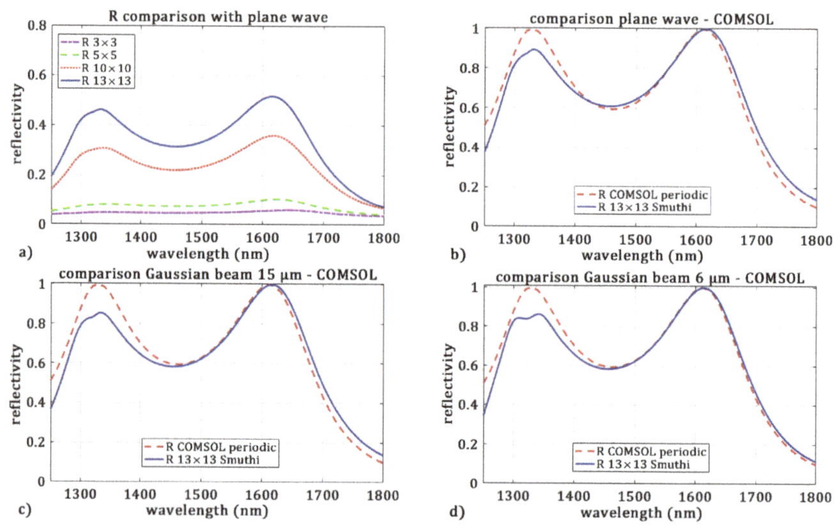

Figure 4. Reflectivity (R) as a function of the wavelength for both the periodic model and the finite model. (**a**) R as a function of the wavelength for the structures obtained by varying the number of cylinders (3×3, 5×5, 10×10, and 13×13), simulated with SMUTHI. The integration surface is the same for all the structures. (**c**,**d**) Comparisons between the two models. For the periodic model with COMSOL, the incident wave is a plane wave; for the space-limited model with SMUTHI there is an incident plane wave (**b**), an incident Gaussian beam 15 µm wide (**c**), and an incident Gaussian beam 6 µm wide (**d**). The values obtained with SMUTHI are normalized so that the maximum is equal to one.

The normalized behaviours are very faithful to the COMSOL behaviours, either with an incident plane wave or with a Gaussian beam. In particular, we see two peaks at 1610 nm and 1330 nm, which are at exactly the same wavelengths for the two models. The resonance at 1610 nm is related to the magnetic dipole inside the cylinder and is also present in the scattering cross section in the isolated structure. The different values around the shorter wavelength peak are due to the finite extension of the structure and the excitation beam; therefore, this feature can only be highlighted through a finite-size model. Indeed, a Gaussian beam can be described by considering a larger amount of k-vectors, which is not possible with the plane-wave excitation employed in COMSOL. This is further evident from Figure 5, where we report also the results obtained considering the flux through the walls around the central cylinder in the finite-size model and the reflectivity of the infinite periodic structure. It is interesting to note that the characteristic of the bigger structure is more similar to the periodic model. As expected, the relative weight of the resonant peak due to the interaction between the cylinders (around 1330 nm) and the single-pillar magnetic dipole (1610 nm) increases with the array size. We emphasize that the resonance at 1610 nm is related to the magnetic dipole in the isolated pillar, thus it is well reproduced even for small arrays and beam sizes. Thus, these results show that it is essential to consider the size of the system and the excitation, to correctly describe a real finite-size system. For this purpose, it should be noted that the size of the structure compared to the wavelength is more important than the number of unit cells considered in the simulation. However, the size of the unit cells plays a key role in determining the interaction strength between the resonators. Indeed, if the cylinders are well separated from each other, they act as isolated sources, and the scattering properties can be evaluated by computing the electromagnetic field of an isolated pillar and considering the system as an array of antennas.

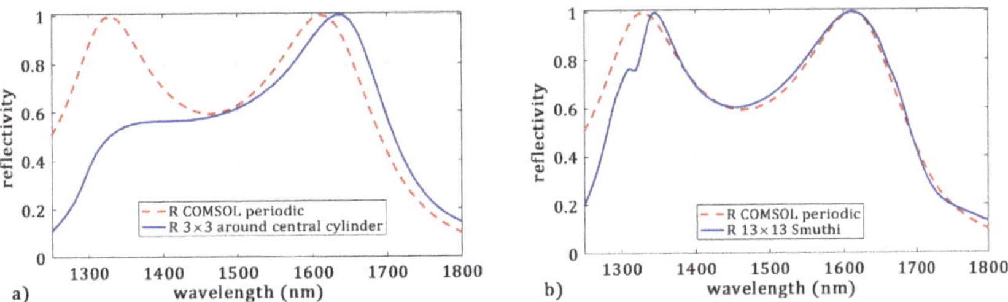

Figure 5. Reflectivity as a function of the wavelength for infinite case (dashed lines) and finite case (continuous lines). For the finite case, the power flux is considered through the walls around the central cylinder for the structures with 3×3 cylinders (**a**) and 13×13 cylinders (**b**). This means that the simulation domain on the xy plane has a width of 856 nm around the central cylinder (compared to the previous value of 15,500 nm). The plots obtained with SMUTHI are normalized to have a maximum equal to one.

To further prove the importance of finite-size effects in conjunction with resonant behaviour, we consider a structure supporting bound states in the continuum. Figure 6 shows the reflectivity as a function of the wavelength for a suspended array of cylinders, which was previously reported for sustaining a BIC [43]. The Gaussian beam waist was kept fixed at 10 μm, while the number of resonators was increased. It is clear that increasing the number of resonators reduces the line width of the Fano resonance associated with the quasi-BIC. We highlight that in [43], the authors introduced a gap in the middle of the cylinders in order to couple to the resonant mode, while in our case this was not necessary since both the illumination and the structure were finite and the modes could couple to the radiating channels.

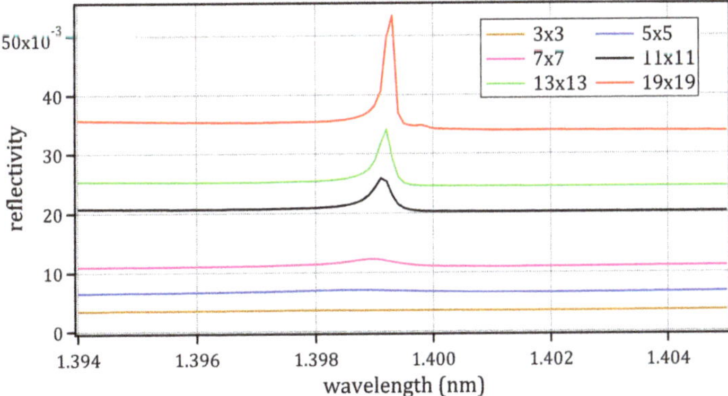

Figure 6. Reflectivity as a function of wavelength for arrays of cylinders (height 563 nm, radius 250 nm, and period 850 nm) with different numbers of elements. The refractive index is 3.49 and the impinging Gaussian beam waist is 10 μm.

5. Conclusions

In conclusion, we built a space-limited model with SMUTHI to simulate a real metasurface, to determine the border effects. First, we tested the software performance in the presence of the metasurface and verified the conservation of energy in the simulated domain ($R + T = 1$), as well as the agreement with the COMSOL simulation. The resulting reflection coefficients R were similar for the two structures, and the produced scattered fields were very similar.

We developed the finite model by introducing an approximation for the scattered field and the refractive indices. Moreover, with the finite-domain model, it was possible to highlight a difference in the higher-frequency peak related to collective excitation. We could also simulate an incident Gaussian beam, which is not possible in the periodic case. SMUTHI allowed us to simulate the wavelength sweep for a large finite structure with 13×13 cylinders, which is impractical to replicate in COMSOL, and also allowed us to simulate the configuration with 3×3 cylinders 5.3 times faster. This approach also allowed us to employ a non-periodic distribution of resonators, which is not possible when exploiting the Floquet theorem. Furthermore, it allowed us to study real situations involving BIC, which are intimately related to the finiteness of the system. The proposed approach may be further extended by analogy with [44], where the authors employed multipolar decomposition to study free-standing single nanoparticles, to simulate nonlinear frequency generation in finite-size metasurfaces.

Author Contributions: Conceptualization, F.M. and A.T.; methodology, L.P.; software, L.P.; investigation, L.C., A.T., F.M. and L.P.; writing—original draft preparation, L.C. and A.T.; writing—review and editing, all the authors contributed equally; supervision, F.F. and C.D.A.; funding acquisition, C.D.A. All authors have read and agreed to the published version of the manuscript.

Funding: This research was funded by Ministero dell'Istruzione, dell'Università e della Ricerca through the PRIN project "NOMEN" (project 2017MP7F8F), the European community through the "METAFAST" project (H2020-FETOPEN-2018-2020, grant agreement no. 899673), and NATO through the Science for Peace and Security (SPS) Programme, project "OPTIMIST" (SPS G5850). L.P. acknowledges Progetto Premiale MIUR "Volume photography" and NVIDIA Corporation for the donation of the Titan X Pascal GPU used for this research.

Acknowledgments: The authors thank Amos Egel for fruitful discussions. The authors are grateful for the financial support of MIUR through the PRIN project "NOMEN" (project 2017MP7F8F), the European community through the "METAFAST" project (H2020-FETOPEN-2018-2020, grant agreement no. 899673), and NATO through the Science for Peace and Security (SPS) Programme, project "OPTIMIST" (SPS G5850). L.P. acknowledges Progetto Premiale MIUR "Volume photography" and NVIDIA Corporation for the donation of the Titan X Pascal GPU used for this research.

Conflicts of Interest: The authors declare no conflict of interest.

Appendix A. Scattering from Axially Symmetric Particles on a Stratified Medium

The SMUTHI software allows the resolution of the electromagnetic scattering problem from an ensemble of axially symmetric particles in a stratified medium. To approach this problem, SMUTHI uses a combination of methods that include the expansion of spherical vector wave functions (SVWFs) to represent the impinging wave in spherical waves, the expansion of plane vector wave functions (PVWFs) to represent the spherical wave in a complete set of plane waves, in order to consider the interaction between the scattered field from the particle and the stratified medium, and a T-matrix approach (or null-field method) with the addition theorem to take into account the interaction between each scatterer. The first step of the method consists of the representation of all the fields as an expansion of SVWFs, then, fixing a scatterer s, they can be written as a superposition of regular harmonic functions for the incoming field or with outgoing harmonic functions [41,42,45–53]:

$$\mathbf{E}_{reg}^{s}(\mathbf{r}) = \sum_{n} a_n^s \mathbf{M}_n^{(1)}(\mathbf{r} - \mathbf{r_s}) \tag{A1}$$

$$\mathbf{E}_{sca}^{s}(\mathbf{r}) = \sum_{n} b_n^s \mathbf{M}_n^{(3)}(\mathbf{r} - \mathbf{r_s}) \tag{A2}$$

where the apexes (1) and (3) represent the types of the SVWFs. In particular, the first type of the Bessel spherical function is used as the harmonic function with (1) and the third type of the same function is used with (3). Each one is centred in the middle of the particle r_s. The second step consists of the simultaneous calculation of the coefficients for the incoming and outgoing fields for each scatterer s. For this, we use the null-field method and the addition theorem, obtaining:

$$b_n^s = \sum_{n'} T_{mn'}^s a_{n'}^s \tag{A3}$$

$$a_n^s = \sum_{s'} \sum_{n'} W_{mn'}^{ss'} b_{n'}^{s'} a_n^{s,in} \tag{A4}$$

having indicated with $T_{mn'}^s$ the T-matrix for a single scatterer, and with the matrix $W_{mn'}^{ss'}$ the mutual interaction between the particles, using the addition theorem and the stratified medium. The third step is characterized by the determination of the interaction between the ensemble of the scattering particles and the stratified medium. The coupling matrix $W_{mn'}^{ss'}$ can be found by considering the superposition of two contributions:

$$W_{mn'}^{ss'} = W_{mn'}^{D,ss'} + W_{mn'}^{R,ss'} \tag{A5}$$

where the first is the coupling term for an infinite medium and the second represents the (multiple) reflections of the wave at the layer interfaces, which are then the incident fields to the particle s. In particular, the last term can be expressed as an expansion of PVWFs, given the flat nature of the interface. Using the Sommerfeld integration form we have:

$$W_{mn'}^{R,ss'} = 4i^{|m'-m|} e^{i|m'-m|} \varphi_{s's} \int_0^\infty \left[\left(f_{nn'}^+(K,z) + f_{nn'}^- K,z) \right) J_{|m'-m|(k\rho)} dK \right] \tag{A6}$$

with

$$f_{nn'}^+(k,z) = \frac{K}{k_z k} \sum_j \left[B_{n,j}^{'+}(L_j^{i_s})_{12} B_{n',j}^- e^{ik_z z} + B_{n,j}^{'-}(L_j^{i_s})_{21} B_{n',j}^+ e^{-ik_z z} \right] \tag{A7}$$

$$f_{nn'}^-(k,z) = \frac{K}{k_z k} \sum_j \left[B_{n,j}^{'+}(L_j^{i_s})_{11} B_{n',j}^- e^{ik_z z} + B_{n,j}^{'-}(L_j^{i_s})_{22} B_{n',j}^+ e^{-ik_z z} \right] \tag{A8}$$

where j indicates the plane wave TE- or TH-polarization, k, K, and k_z are the wavenumber, the wave vector's radial cylinder and the z component in the scattering particle surface, respectively, $B_{n,j}^{'\pm}$ are the transformation coefficients in matrix form between the SVWFs and PVWFs, and $L_j^{i_s}$ is a generalized reflection coefficient in matrix form encoding the overall response of the flat layer interfaces. A complete derivation of the values can be found in [41,42,45–53].

References

1. Kuznetsov, A.I.; Miroshnichenko, A.E.; Brongersma, M.L.; Kivshar, Y.S.; Luk'yanchuk, B. Optically resonant dielectric nanostructures. *Science* **2016**, *354*, aag2472. [CrossRef] [PubMed]
2. Carletti, L.; Locatelli, A.; Stepanenko, O.; Leo, G.; De Angelis, C. Enhanced second-harmonic generation from magnetic resonance in AlGaAs nanoantennas. *Opt. Express* **2015**, *23*, 26544. [CrossRef] [PubMed]
3. Gili, V.F.; Carletti, L.; Locatelli, A.; Rocco, D.; Finazzi, M.; Ghirardini, L.; Favero, I.; Gomez, C.; Lemaître, A.; Celebrano, M.; et al. Monolithic AlGaAs second-harmonic nanoantennas. *Opt. Express* **2016**, *24*, 15965. [CrossRef] [PubMed]
4. Liu, S.; Sinclair, M.B.; Saravi, S.; Keeler, G.A.; Yang, Y.; Reno, J.; Peake, G.M.; Setzpfandt, F.; Staude, I.; Pertsch, T.; et al. Resonantly Enhanced Second-Harmonic Generation Using III-V Semiconductor All-Dielectric Metasurfaces. *Nano Lett.* **2016**, *16*, 5426–5432. [CrossRef]
5. Ghirardini, L.; Carletti, L.; Gili, V.; Pellegrini, G.; Duò, L.; Finazzi, M.; Rocco, D.; Locatelli, A.; De Angelis, C.; Favero, I.; et al. Polarization properties of second-harmonic generation in AlGaAs optical nanoantennas. *Opt. Lett.* **2017**, *42*, 559–562. [CrossRef]

6. Marino, G.; Solntsev, A.S.; Xu, L.; Gili, V.; Carletti, L.; Poddubny, A.N.; Smirnova, D.; Chen, H.; Zhang, G.; Zayats, A.; et al. Sum-frequency generation and photon-pair creation in AlGaAs nano-scale resonators. In Proceedings of the 2017 Conference on Lasers and Electro-Optics (CLEO), San Jose, CA, USA, 14–19 May 2017.

7. Smirnova, D.; Kivshar, Y.S. Multipolar nonlinear nanophotonics. *Optica* **2016**, *3*, 1241–1255. doi: 10.1364/optica.3.001241. [CrossRef]

8. Lapine, M. New degrees of freedom in nonlinear metamaterials. *Phys. Status Solidi (B) Basic Res.* **2017**, *254*, 1600462. [CrossRef]

9. Krasnok, A.; Tymchenko, M.; Alù, A. Nonlinear metasurfaces: A paradigm shift in nonlinear optics. *Mater. Today* **2018**, *21*, 8–21.

10. Heydari, S.; Bazgir, M.; Zarrabi, F.B.; Gandji, N.P.; Rastan, I. Novel optical polarizer design based on metasurface nano aperture for biological sensing in mid-infrared regime. *Opt. Quantum Electron.* **2017**, *49*, 83. [CrossRef]

11. Ebrahimi, S.; Poorgholam-Khanjari, S. Highly Q-factor optical metasurface based on DNA nanorods with Fano response for on-chip optical sensing. *Optik* **2021**, *232*, 166576.: 10.1016/j.ijleo.2021.166576. [CrossRef]

12. Mobasser, S.; Poorgholam-Khanjari, S.; Bazgir, M.; Zarrabi, F.B. Highly Sensitive Reconfigurable Plasmonic Metasurface with Dual-Band Response for Optical Sensing and Switching in the Mid-Infrared Spectrum. *J. Electron. Mater.* **2021**, *50*, 120–128. [CrossRef]

13. Krasnok, A.; Savelev, R. All-Dielectric Nanophotonics: Fundamentals , Fabrication and Applications. In *World Scientific Handbook of Metamaterials and Plasmonics-Volume 1*; Mayer, S., Ed.; World Scientific: Singapore, 2017; Chapter 8, pp. 337–385.

14. Jahani, S.; Jacob, Z. All-dielectric metamaterials. *Nat. Nanotechnol.* **2016**, *11*, 23–36. [CrossRef] [PubMed]

15. Zhao, Q.; Zhou, J.; Zhang, F.; Lippens, D. Mie resonance-based dielectric metamaterials. *Mater. Today* **2009**, *12*, 60–69. [CrossRef]

16. Chen, H.T.; Taylor, A.J.; Yu, N. A review of metasurfaces: Physics and applications. *Rep. Prog. Phys.* **2016**, *79*.

17. Carletti, L.; Rocco, D.; Locatelli, A.; De Angelis, C.; Gili, V.F.; Ravaro, M.; Favero, I.; Leo, G.; Finazzi, M.; Ghirardini, L.; et al. Controlling second-harmonic generation at the nanoscale with monolithic AlGaAs-on-AlOx antennas. *Nanotechnology* **2017**, *28*, 114005. [CrossRef]

18. Carletti, L.; Marino, G.; Ghirardini, L.; Gili, V.F.; Rocco, D.; Favero, I.; Locatelli, A.; Zayats, A.V.; Celebrano, M.; Finazzi, M.; et al. Nonlinear Goniometry by Second-Harmonic Generation in AlGaAs Nanoantennas. *ACS Photonics* **2018**, *5*, 4386–4392. [CrossRef]

19. Carletti, L.; Locatelli, A.; Neshev, D.; De Angelis, C. Shaping the Radiation Pattern of Second-Harmonic Generation from AlGaAs Dielectric Nanoantennas. *ACS Photonics* **2016**, *3*, 1500–1507. [CrossRef]

20. Camacho-Morales, R.; Rahmani, M.; Kruk, S.; Wang, L.; Xu, L.; Smirnova, D.A.; Solntsev, A.S.; Miroshnichenko, A.; Tan, H.H.; Karouta, F.; et al. Nonlinear Generation of Vector Beams from AlGaAs Nanoantennas. *Nano Lett.* **2016**, *16*, 7191–7197. [CrossRef]

21. Kruk, S.S.; Camacho-Morales, R.; Xu, L.; Rahmani, M.; Smirnova, D.A.; Wang, L.; Tan, H.H.; Jagadish, C.; Neshev, D.N.; Kivshar, Y.S. Nonlinear optical magnetism revealed by second-harmonic generation in nanoantennas. *Nano Lett.* **2017**, *17*, 3914–3918. [CrossRef]

22. Shcherbakov, M.R.; Vabishchevich, P.P.; Shorokhov, A.S.; Chong, K.E.; Choi, D.Y.; Staude, I.; Miroshnichenko, A.E.; Neshev, D.N.; Fedyanin, A.A.; Kivshar, Y.S. Ultrafast All-Optical Switching with Magnetic Resonances in Nonlinear Dielectric Nanostructures. *Nano Lett.* **2015**, *15*, 6985–6990. [CrossRef]

23. Della Valle, G.; Hopkins, B.; Ganzer, L.; Stoll, T.; Rahmani, M.; Longhi, S.; Kivshar, Y.S.; De Angelis, C.; Neshev, D.N.; Cerullo, G. Nonlinear Anisotropic Dielectric Metasurfaces for Ultrafast Nanophotonics. *ACS Photonics* **2017**, *4*, 2129–2136. [CrossRef]

24. Shcherbakov, M.R.; Neshev, D.N.; Hopkins, B.; Shorokhov, A.S.; Staude, I.; Melik-Gaykazyan, E.V.; Decker, M.; Ezhov, A.A.; Miroshnichenko, A.E.; Brener, I.; et al. Enhanced third-harmonic generation in silicon nanoparticles driven by magnetic response. *Nano Lett.* **2014**, *14*, 6488–6492. [CrossRef] [PubMed]

25. Yang, Y.; Wang, W.; Boulesbaa, A.; Kravchenko, I.I.; Briggs, D.P.; Puretzky, A.; Geohegan, D.; Valentine, J. Nonlinear Fano-Resonant Dielectric Metasurfaces. *Nano Lett.* **2015**, *15*, 7388–7393. [CrossRef] [PubMed]

26. Babicheva, V.E.; Evlyukhin, A.B. Multipole lattice effects in high refractive index metasurfaces. *J. Appl. Phys.* **2021**, *129*, 040902.

27. Castellanos, G.W.; Bai, P.; Gómez Rivas, J. Lattice resonances in dielectric metasurfaces. *J. Appl. Phys.* **2019**, *125*, 213105.

28. Utyushev, A.D.; Zakomirnyi, V.I.; Ershov, A.E.; Gerasimov, V.S.; Karpov, S.V.; Rasskazov, I.L. Collective Lattice Resonances in All-Dielectric Nanostructures under Oblique Incidence. *Photonics* **2020**, *7*, 24. [CrossRef]

29. Joseph, S.; Pandey, S.; Sarkar, S.; Joseph, J. Bound states in the continuum in resonant nanostructures: An overview of engineered materials for tailored applications. *Nanophotonics* **2021**, *10*, 4175–4207.:10.1515/nanoph-2021-0387. [CrossRef]

30. Zundel, L.; Manjavacas, A. Finite-size effects on periodic arrays of nanostructures. *J. Phys. Photonics* **2018**, *1*, 015004.

31. Zakomirnyi, V.I.; Ershov, A.E.; Gerasimov, V.S.; Karpov, S.V.; Ågren, H.; Rasskazov, I.L. Collective lattice resonances in arrays of dielectric nanoparticles: a matter of size. *Opt. Lett.* **2019**, *44*, 5743–5746. [CrossRef]

32. Zakomirnyi, V.I.; Karpov, S.V.; Ågren, H.; Rasskazov, I.L. Collective lattice resonances in disordered and quasi-random all-dielectric metasurfaces. *J. Opt. Soc. Am. B* **2019**, *36*, E21–E29. [CrossRef]

33. Ziolkowski, R.W. Mixtures of Multipoles—Should They Be in Your EM Toolbox? *IEEE Open J. Antennas Propag.* **2022**, *3*, 154–188. [CrossRef]

34. Czajkowski, K.M.; Bancerek, M.; Antosiewicz, T.J. Multipole analysis of substrate-supported dielectric nanoresonator metasurfaces via the *T*-matrix method. *Phys. Rev. B* **2020**, *102*, 085431. [CrossRef]

35. Li, J.; Verellen, N.; Van Dorpe, P. Engineering electric and magnetic dipole coupling in arrays of dielectric nanoparticles. *J. Appl. Phys.* **2018**, *123*, 083101.

36. Babicheva, V.E.; Evlyukhin, A.B. Analytical model of resonant electromagnetic dipole-quadrupole coupling in nanoparticle arrays. *Phys. Rev. B* **2019**, *99*, 195444. [CrossRef]
37. Theobald, D.; Beutel, D.; Borgmann, L.; Mescher, H.; Gomard, G.; Rockstuhl, C.; Lemmer, U. Simulation of light scattering in large, disordered nanostructures using a periodic T-matrix method. *J. Quant. Spectrosc. Radiat. Transf.* **2021**, *272*, 107802. [CrossRef]
38. Kostyukov, A.S.; Rasskazov, I.L.; Gerasimov, V.S.; Polyutov, S.P.; Karpov, S.V.; Ershov, A.E. Multipolar Lattice Resonances in Plasmonic Finite-Size Metasurfaces. *Photonics* **2021**, *8*, 109. [CrossRef]
39. Egel, A.; Czajkowski, K.M.; Theobald, D.; Ladutenko, K.; Kuznetsov, A.S.; Pattelli, L. SMUTHI: A python package for the simulation of light scattering by multiple particles near or between planar interfaces. *J. Quant. Spectrosc. Radiat. Transf.* **2021**, *273*, 107846.: 10.1016/j.jqsrt.2021.107846. [CrossRef]
40. Sharma, K.K. *Optics: Principles and Applications*, 1st ed.; Academic Press: Cambridge, MA, USA, 2006.
41. Egel, A.; Kettlitz, S.W.; Lemmer, U. Efficient evaluation of Sommerfeld integrals for the optical simulation of many scattering particles in planarly layered media. *J. Opt. Soc. Am. A* **2016**, *33*, 698–706. [CrossRef]
42. Egel, A.; Lemmer, U. Dipole emission in stratified media with multiple spherical scatterers: Enhanced outcoupling from OLEDs. *J. Quant. Spectrosc. Radiat. Transf.* **2014**, *148*, 165–176. [CrossRef]
43. Tognazzi, A.; Rocco, D.; Gandolfi, M.; Locatelli, A.; Carletti, L.; De Angelis, C. High Quality Factor Silicon Membrane Metasurface for Intensity-Based Refractive Index Sensing. *Optics* **2021**, *2*, 193–199. [CrossRef]
44. Smirnova, D.A.; Khanikaev, A.B.; Smirnov, L.A.; Kivshar, Y.S. Multipolar Third-Harmonic Generation Driven by Optically Induced Magnetic Resonances. *ACS Photonics* **2016**, *3*, 1468–1476.
45. Videen, G. Light scattering from a particle on or near a perfectly conducting surface. *Opt. Commun.* **1995**, *115*, 1–7. [CrossRef]
46. Wriedt, T.; Doicu, A. Light scattering from a particle on or near a surface. *Opt. Commun.* **1998**, *152*, 376–384. [CrossRef]
47. Tsang, L.; Kong, J.A.; Ding, K.H. *Scattering of Electromagnetic Waves. Theory and Applications*, 1st ed.; Wiley: Hoboken, NJ, USA, 2000.
48. Doicu, A.; Wriedt, T.; Eremin, Y.A. *Light Scattering by Systems of Particles*, 1st ed.; Springer: Berlin/Heidelberg, Germany, 2006.
49. Frezza, F.; Mangini, F.; Pajewski, L.; Schettini, G.; Tedeschi, N. Spectral domain method for the electromagnetic scattering by a buried sphere. *J. Opt. Soc. Am. A* **2013**, *30*, 783–790. [CrossRef] [PubMed]
50. Frezza, F.; Mangini, F.; Tedeschi, N. Electromagnetic scattering by two concentric spheres buried in a stratified material. *J. Opt. Soc. Am. A* **2015**, *32*, 277–286. [CrossRef]
51. Frezza, F.; Mangini, F. Vectorial spherical-harmonics representation of an inhomogeneous elliptically polarized plane wave. *J. Opt. Soc. Am. A* **2015**, *32*, 1379–1383. [CrossRef]
52. Frezza, F.; Mangini, F. Electromagnetic scattering by a buried sphere in a lossy medium of an inhomogeneous plane wave at arbitrary incidence: Spectral-domain method. *J. Opt. Soc. Am. A* **2016**, *33*, 947–953. [CrossRef]
53. Frezza, F.; Mangini, F. Electromagnetic scattering of an inhomogeneous elliptically polarized plane wave by a multilayered sphere. *J. Electromagn. Waves Appl.* **2016**, *30*, 492–504. [CrossRef]

Article

Investigations on Field Distribution along the Earth's Surface of a Submerged Line Current Source Working at Extremely Low Frequency Band

Ke Yang [1,2,*], Jinhong Wang [1,2], Shuwen Liu [1], Kai Ding [3], Hao Li [3] and Bin Li [1,2]

1 School of Marine Science and Technology, Northwestern Polytechnical University, Xi'an 710072, China; jhwang@nwpu.edu.cn (J.W.); sw_liu@nwpu.edu.cn (S.L.); libin_cme@nwpu.edu.cn (B.L.)
2 Key Laboratory of Marine Acoustics Information Perception, Ministry of Industry and Information Technology, Xi'an 710072, China
3 Science and Technology on Near-Surface Detection Laboratory, Wuxi 214182, China; k.ding@126.com (K.D.); hli@126.com (H.L.)
* Correspondence: k.yang@nwpu.edu.cn

Abstract: A numerical analysis on field distribution along the Earth's surface of a line current source submerged in the ground is conducted in this paper to investigate the potential of the extremely low frequency (ELF) technology in the envisioned long-distance communication techniques. The problem is modeled as a submerged horizontal electric dipole (HED) in a two-layered homogeneous half space and solved by the combined numerical methods of the Romberg-Euler method and Gauss-Laguerre method. The model is validated by experimental results with only a maximum 10% error at 9 Hz around 490 m. Meanwhile, the study shows that the ELF signals emitted by a submerged line current source can transmit at least 1 km with a current sensor sensitivity of 0.1 pT. These results indicate the possibility of applying of ELF technology to long-distance communication or the long-distance transmedia detection.

Keywords: electromagnetic wave distribution; extremely low frequency; layered-space problem; horizontal electric dipole

Citation: Yang, K.; Wang, J.; Liu, S.; Ding K.; Li, H.; Li, B. Investigations on Field Distribution along the Earth's Surface of a Submerged Line Current Source Working at Extremely Low Frequency Band. *Electronics* **2022**, *11*, 1116. https://doi.org/10.3390/electronics11071116

Academic Editors: Giulio Antonini, Daniele Romano and Luigi Lombardi

Received: 9 December 2021
Accepted: 18 March 2022
Published: 1 April 2022

Publisher's Note: MDPI stays neutral with regard to jurisdictional claims in published maps and institutional affiliations.

1. Introduction

The investigation on extremely low frequency electromagnetic wave distribution has been a classical problem because of the wide applications of technologies in marine engineering such as undersea detection, undersea sensor networks, coastal surveillance, seabed communication, etc. The problem has attracted increasing attention and investigations, which are mainly focused on field distribution in stratified media since the first analysis for the electromagnetic wave along the planar boundary as early as 1907 [1]. Then, Dr. Sommerfeld investigated the excited field of a dipole in the multilayered lossy media where the concept of Sommerfeld integrals [2] is proposed, and a field theory in a homogeneous half-space is fully developed in [3]. Since the attenuation rates of ELF waves could be as low as 1 dB/1000 km, stated in reference [4], the potential of the applications of such technologies to long-distant communication has gained researcher's attention. Currently, numerous investigations have been conducted on the ELF EM wave propagation based on an isotropic conductivity Earth model where field distribution by a dipole submerged in a lossy media is thoroughly studied [5–9]. Chave A. D. studies undersea communication by free propagation of ULF/ELF electromagnetic waves from harmonic dipole sources located on or near the sea floor [10]. In 2017, the scenario of a three-layered conducting media including air, seawater and seabed has been discussed [11], where the ELF electromagnetic fields generated by a submerged horizontal electric dipole in the shallow sea are fully investigated. Later, the studies have been extended to the detection of underwater moving objects [12,13]. In 2018, the seabed environment is treated as a three-layered media and the field distribution is numerically investigated [14]. At the same time, the ELF EM signal has widely used in drilling system to carry downhole measurement data where the studies of channel models and demodulation techniques are essential [15]. Meanwhile, the existence of Zenneck surface wave has drawn researchers' attention, which was proposed by Baño [3]. Successive studies have been conducted by Dr. King on lateral electromagnetic waves in 1980s while the work is extended from two-layered to three-layered media by Dr. Dunn [16–19].

It can be easily observed that large amounts of studies have been conducted on theoretical analysis and numerical modeling, but there is scarcity of fulfillment of the transmission of ELF signals [20–22]. Furthermore, the majority of current measurements are limited to the propagation of ELF wave in the sea while the work on wave propagation along Earth's surfaces remains vacant. To fill these gaps, an experiment is first designed with a line current source submerged in the ground where the emitted magnetic field signals along the mountain surface are collected by the commercial magnetic sensor. Then, the submerged line current source is modeled as a submerged horizontal electric dipole, and the field distribution is calculated by a combined method of Romberg-Euler method and Gauss-Laguerre method. Furthermore, numerical results are compared with experimental ones to validate the models and investigate field performances.

The remainder of the paper is organized as follows. Section 2 discusses the theoretical model of the line current source while the experiment's descriptions are illustrated in Section 3. At the same time, the performance of the numerical model is also investigated in Section 3. In the end, a brief conclusion is drawn in Section 4.

2. Theoretical Model

To save the space and to enhance performance, a horizontal line current is adopted in our study, which is approximated as a horizontal electric dipole source theoretically. Not losing generality, an x-directed horizontal electric dipole is investigated, as shown in Figure 1, which is placed at point $z = d$ on the downward-directed z-axis in the half-space where $z \geq 0$, which is region 1. Region 2 is the space where $z \leq 0$. The wave number of each region is $k_j = \omega \sqrt{\mu_j \epsilon_j}$, where $\epsilon_j = \epsilon_0(\epsilon_{rj} - j\sigma_i/\omega\epsilon_0)$, $\mu_j = \mu_{ri}\mu_0$ and $j = 1,2$. ϵ_0 and μ_0 are the permittivity and permeability of free space. Since soil and rock are all non-magnetic, $\mu_{r1} = \mu_{r2} = 1$.

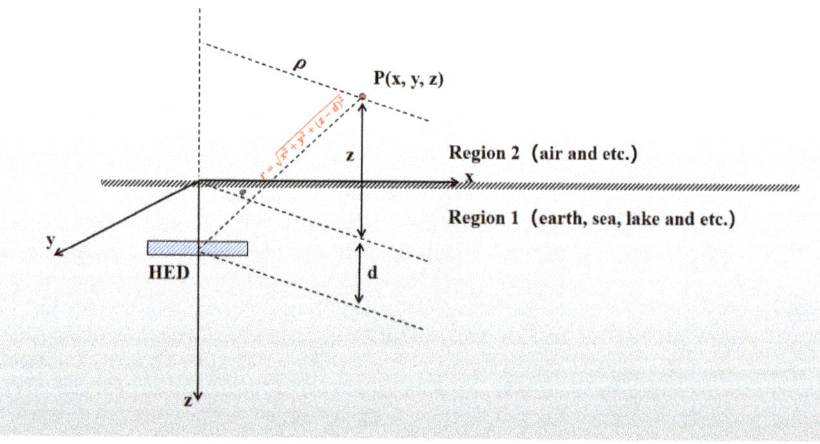

Figure 1. Schematics of x-directed horizontal electric dipole at $(0,0,d)$ and measuring point P at (x,y,z) or (ρ,ϕ,z).

Therefore, the problem can be described as follows. Maxwell's equation in the two region are as follows:

$$\begin{aligned}
\nabla \times \mathbf{E}_j &= i\omega \mathbf{B}_j \\
\nabla \times \mathbf{B}_j &= \mu_0(-i\omega\epsilon_j \mathbf{E}_j + \mathbf{J})
\end{aligned} \tag{1}$$

where $\mathbf{J} = \hat{x} Il\delta(x)\delta(y)\delta(z-d)$ is the electric moment.

$\mathbf{E}_j$ and $\mathbf{B}_j$ will be solved with the following boundary conditions.

$$\begin{aligned}
E_{1x}(x,y,0) &= E_{2x}(x,y,0); \\
E_{1y}(x,y,0) &= E_{2y}(x,y,0); \\
k_1^2 E_{1z}(x,y,0) &= k_2^2 E_{2z}(x,y,0); \\
\mathbf{B}_1(x,y,0) &= \mathbf{B}_2(x,y,0)
\end{aligned} \tag{2}$$

To investigate the performance of the wave, we usually measure the fields in region 2. Thus, the field components in it are solved and illustrated below [23]:

$$\begin{aligned}
E_{2\rho} &= -\frac{Il\omega\mu_0}{4\pi}\cos\phi \int_0^\infty \left(\frac{[J_0(\lambda\rho) + J_2(\lambda\rho)]}{M} + \frac{\gamma_1\gamma_2[J_0(\lambda\rho) - J_2(\lambda\rho)]}{N}\right)e^{i(\gamma_1 d - \gamma_2 z)}\lambda d\lambda \\
E_{2\phi} &= \frac{Il\omega\mu_0}{4\pi}\sin\phi \int_0^\infty \left(\frac{[J_0(\lambda\rho) - J_2(\lambda\rho)]}{M} + \frac{\gamma_1\gamma_2[J_0(\lambda\rho) + J_2(\lambda\rho)]}{N}\right)e^{i(\gamma_1 d - \gamma_2 z)}\lambda d\lambda \\
E_{2z} &= -\frac{Il\omega\mu_0}{2\pi}\cos\phi \int_0^\infty \frac{\gamma_1 J_1(\lambda\rho)e^{i(\gamma_1 d - \gamma_2 z)}\lambda^2}{N}d\lambda \\
B_{2\rho} &= \frac{Il\mu_0}{4\pi}\sin\phi \int_0^\infty \left(\frac{\gamma_2[J_0(\lambda\rho) - J_2(\lambda\rho)]}{M} + \frac{k_2^2\gamma_1[J_0(\lambda\rho) + J_2(\lambda\rho)]}{N}\right)e^{i(\gamma_1 d - \gamma_2 z)}\lambda d\lambda \\
B_{2\phi} &= \frac{Il\mu_0}{4\pi}\cos\phi \int_0^\infty \left(\frac{\gamma_2[J_0(\lambda\rho) + J_2(\lambda\rho)]}{M} + \frac{k_2^2\gamma_1[J_0(\lambda\rho) - J_2(\lambda\rho)]}{N}\right)e^{i(\gamma_1 d - \gamma_2 z)}\lambda d\lambda \\
B_{2z} &= \frac{iIl\mu_0}{2\pi}\sin\phi \int_0^\infty \frac{J_1(\lambda\rho)e^{i(\gamma_1 d - \gamma_2 z)}\lambda^2}{M}d\lambda
\end{aligned} \tag{3}$$

where $M \equiv \gamma_1 + \gamma_2$, $N \equiv k_2^2\gamma_2 + k_1^2\gamma_1$ and J_n is the Bessel function. $\gamma_j^2 = k_j^2 - \xi^2 - \eta^2$, which is defined by the translational invariance $\mathbf{E}(x,y,z) = \frac{1}{(2\pi)^2}\int_{-\infty}^{\infty} d\xi \int_{-\infty}^{\infty} d\eta e \mathbf{E}(\xi,\eta,z)$ [23].

The calculation of Equation (3) would require the analysis of the Sommerfeld Integral, for which its general form is described as follows:

$$S = \int_0^\infty f(\lambda)J_n(\lambda\rho)e^{-g(\lambda)}d\lambda \tag{4}$$

where $f(\lambda)$ and $g(\lambda)$ are the complex function of the frequency and medium properties. $J_n(\lambda\rho)$ is the first kind Bessel function of order n. To evaluate SI, numerous poles would appear in the integral path, which would make the numerical calculation divergent. To solve such problems, numerous techniques have been developed and a combined numerical method with the Gauss-Laguerre method and Romberg–Euler method is applied in the paper based on the current studies [24]. A brief introduction of both methods is illustrated as follows.

2.1. Gauss-Laguerre Method

The Gauss-Laguerre quadrature method can be expressed as follows:

$$\int_0^\infty f(x)e^{-x}dx \approx \sum_{i=0}^{n} A_i f(x_i) \tag{5}$$

where x_i is the ith null point of the Laguerre polynomial of order $N+1$, and w_i is the corresponding weight; they can be easily obtained from the relevant literature. The approximation in Equation (5) is exact when $f(x)$ is a polynomial of a degree less than $2N+1$; thus, the Guass–Laguerre quadrature succeeds if $f(x)$ can be adequately represented by a polynomial of finite order for range $[0,\infty]$. It is obvious that the higher order Laguerre polynomial can produce more accurate integral results but involves more computation time. For example, in Equation (3), by permitting the following, B_{2z} can be evaluated by Equation (5).

$$f(x) = \frac{J_1(x\rho)e^{i(\gamma_1 d - \gamma_2 z)x^2}e^x}{M} \tag{6}$$

Although more complicated than B_{2z}, the other components can also be evaluated in this manner.

This method is suitable for the case where the horizontal distance is less than the vertical distance, i.e., $\rho \leq z$.

2.2. Romberg-Euler Method

The Romberg–Euler composite method can by expressed as follows:

$$S = \int_0^\infty W(\lambda)J_n(\lambda\rho)e^{-g(\lambda)}d\lambda \approx \sum_{m=1}^\infty \int_{x_{m-1}}^{x_m} W(\lambda)J_n(\lambda\rho)e^{-g(\lambda)}d\lambda = \sum_{m=1}^\infty B_m = S_p \tag{7}$$

where $x_0 = 0$, and x_m is the first root of the Bessel function $J_n(\lambda\rho)$. The value of the integral S is the limit to which sequence S_p converges.

The termination condition of the iteration is stated as follows.

$$\left|\frac{\Delta S_p}{S_p}\right| = \left|\frac{(S_p - S_{p-1})}{S_p}\right| = \left|\frac{B_p}{S_p}\right| < \epsilon \tag{8}$$

Take B_z in Equation (3) as an example again; here, we can permit $W(\lambda) = \frac{1}{M}$ and $g(\lambda) = -i(\gamma_1 d - \gamma_2 z)\lambda^2$. Then, B_{2z} can be evaluated by Equation (7).

The integration is performed between the zeros of the Bessel function, and the truncated infinite integral is expressed as a sum of integral between successive Bessel zeros. Since the envelope of $W(\lambda)$ decays more slowly than the rapidly oscillating Bessel function, especially for large values of the argument, the Euler transform can be used to speed up the converge of the piecewise integration sequence. The Romberg quadrature method is adopted for piecewise integration in B_m.

To evaluate the performance of the above methods, the calculation times are usually compared with different values of $\frac{\rho}{z}$. Studies shows that when $\rho > z$, the Romberg–Euler method outperforms the Gauss-Laguerre one [24] and when $\rho < z$, Gauss-Laguerre method performs better.

3. Experiments and Results Discussions

3.1. Experiments Description

An experiment has been conducted in a reservoir in the Qin Mountain in Northwestern China (109°49′ E, 33°42′ N), shown in Figure 2a, to study the radiation performances of the submerged line current and the wave diffusion along the Earth's surface. The soil of the region was composed of mud, sand and crushed granite where conductivity is around 5 S/m. The line current source is buried around 1-m deep in a shallow horizontal layer and connected to a transmitter which would send ELF signals. The schematics of the source are shown in Figure 2b. In the experiment, the copper wire is around 200 m long, the radius is around 1 mm and the emission current is constantly 0.07 A. Thus, the emission intensity of the envisioned electric dipole is 14 Am. The emitting frequencies of the ELF signals are set at 3 Hz, 5 Hz and 9 Hz.

(**a**) Topographic map of the experiment layout

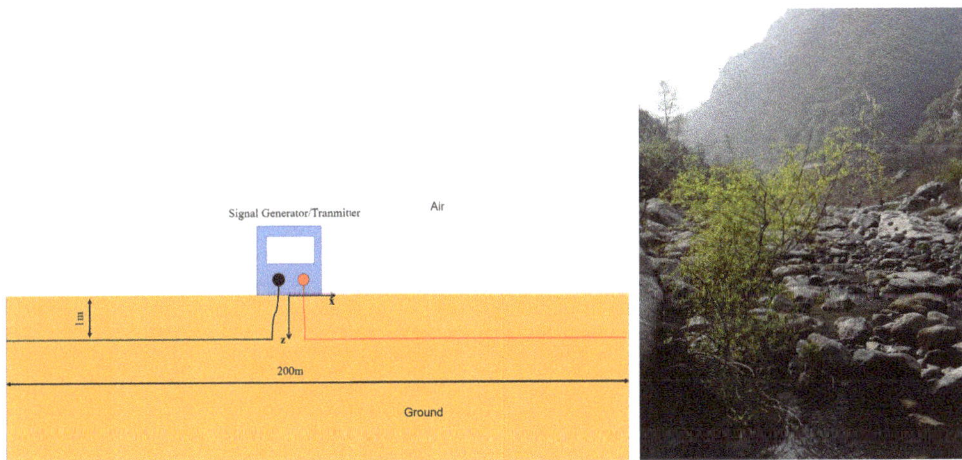

(**b**) Schematic layout of source

(**c**) Source Environment along Hao Creek

Figure 2. Schematics of the experiment setup.

A commercial magnetic sensor produced by Huashun Ltd. (Xi'an, China) with a high sensitivity of 216 mV/nT and low noise level of 0.057 pT/$\sqrt{\text{Hz}}$ at 1 Hz, shown in Figure 3a, is adopted to collect magnetic signals. The details of the ferrite rod sensor is shown in Table 1. To investigate the magnetic fields' complete performance, three sensors were perpendicularly assembled relative each other, shown in Figure 3b, which are placed as high as 20 cm away from the Earth's surface. The three-axis magnetic collection system would transfer the collecting magnetic field into voltage with 24-bit sampling, which would be recorded in the data acquisition system based on NI modules with sampling rates of 200 Hz, as shown in Figure 3c. To obtain the field strength, the values at the frequency of interest would be collected and transferred by sensitivity. A 5 Hz square-wave signal collected by the system is shown in Figures 4 and 5.

To check the source states, a reference point is fixed closely to the line current source, which records the received magnetic signal at 50 m. Other spots at 490 m, 632 m and 735 m away from the source are chosen according to the environment in the mountain. Figure 3c shows the receiver placed in a 5-m deep cave at an distance of 490 m.

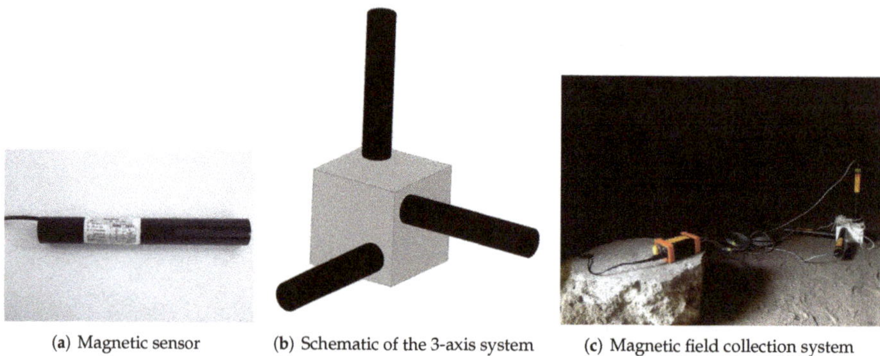

Figure 3. Commercial sensors to collect magnetic signals.

Table 1. Parameters of the ferrite rod sensor.

Paramters	Values
Bandwith	0.1 Hz–200 Hz
Noise Level	$0.057\,\mathrm{pT}/\sqrt{\mathrm{Hz}}@1\,\mathrm{Hz}$
Measurement Range	$\pm10\,\mathrm{nT}$
Output Voltage	$\pm10\,\mathrm{V}$
NRP	$\leq0.35\,\mathrm{W}$

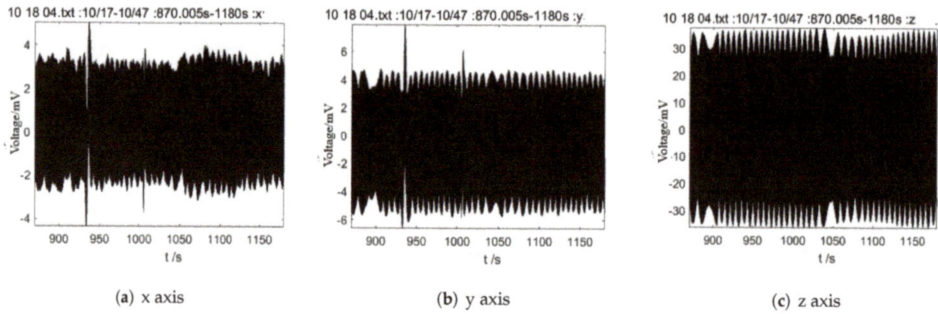

Figure 4. Original signal of a 5 Hz square wave collected by the system.

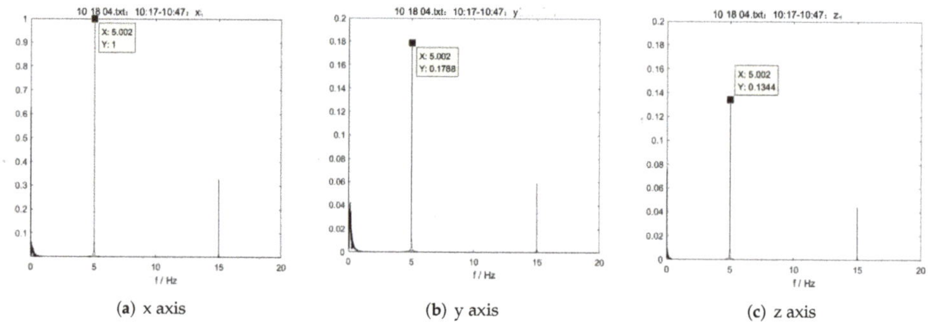

Figure 5. Post-processed measured values vs. frequency of a 5 Hz square signal.

3.2. Results Discussion

3.2.1. Validation of the Numerical Model

To investigate the performance of the proposed numerical model and the combined method, the measured magnetic fields strengths are compared with the numerical results of the submerged HED model. As shown in Figure 1, the source is selected as an HED with the electric moment of 14 Am, and it was located in a two-half homogeneous space. The HED is 1 m away from the interface and located in region 1 where $\epsilon_r = 10, \mu_r = 1, \sigma = 5\,\text{S/m}$, which are the general parameters of the soil. Region 2 would be air with $\epsilon_r = 1, \mu_r = 1, \sigma = 0\,\text{S/m}$, and the fields of interest are analyzed.

The comparison of the two field strengths is illustrated in Figure 6 with distance as the horizontal axis. From the figure, we can see that both data agree well but with some slight differences with 0.026 nT as the maximum (10%) at 9 Hz around 490 m. The errors are mainly caused by the ideal assumption of the ideal layered structures and the noise produced by the collection system and introduced by the environment. Thus, it can be easily concluded that the proposed model would be a promising substitute for the preliminary study of such channels at the ELF band, especially for locations that are hard to reach. At the same time, it can be easily observed that, in the ELF band, the influences of the frequency on field strength are limited, especially in the closer range.

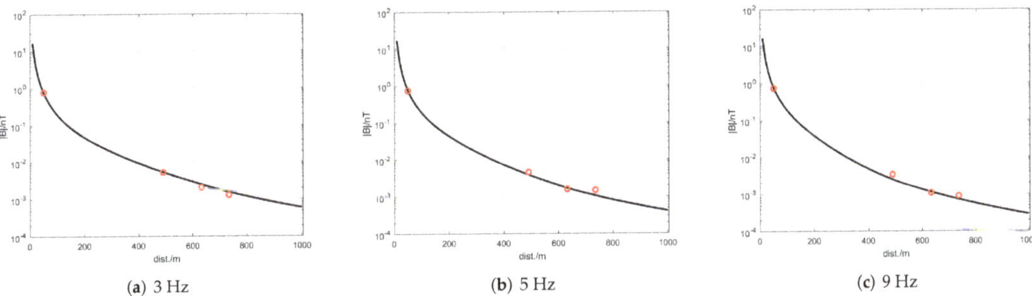

(a) 3 Hz   (b) 5 Hz   (c) 9 Hz

Figure 6. Comparison of measured data with the theoretical values for ELF waves at 3 Hz, 5 Hz and 9 Hz.

3.2.2. Investigations on the Field Distribution

From the previous study, it can be clearly concluded that a magnetic ELF signals could be produced by a submerged line current and such signals can travel along the Earth's surface. Although the signal strengths are at the level of nT, magnetic signals can be collected at least 1 km for all ELF bands up to 9 Hz due to the development of the current sensor technology, which would increase sensitivity to the level of $sub\text{-}pT$.

The electric field and magnetic field distributions at *xoy-plane* ($z = 1$ m) of the ELF wave at 5 Hz with an x-axis located HED as the source are shown in Figure 7. The electric moments is 14 Am, and the figure is illustrated in logarithm form. A typical distribution of an electric dipole can be observed and the magnetic field suffers less power loss than the electric field, which indicates the potential of the magnetic signals of ELF communication techniques.

The power loss of the signals is illustrated in Figure 8. To calculate path loss, the magnitude of the magnetic field was measured and recorded at 50 m, which was chosen as reference B_0. Then, the ratio of the field magnitude at ith point B_i to the ones at reference point B_0 would be calculated and followed by a logarithmic operation, i.e., $PL = 20lg\frac{B_i}{B_0}$. From the figure, it can be observed that path loss increases with an increase in the distance, but frequency would have limited impact.

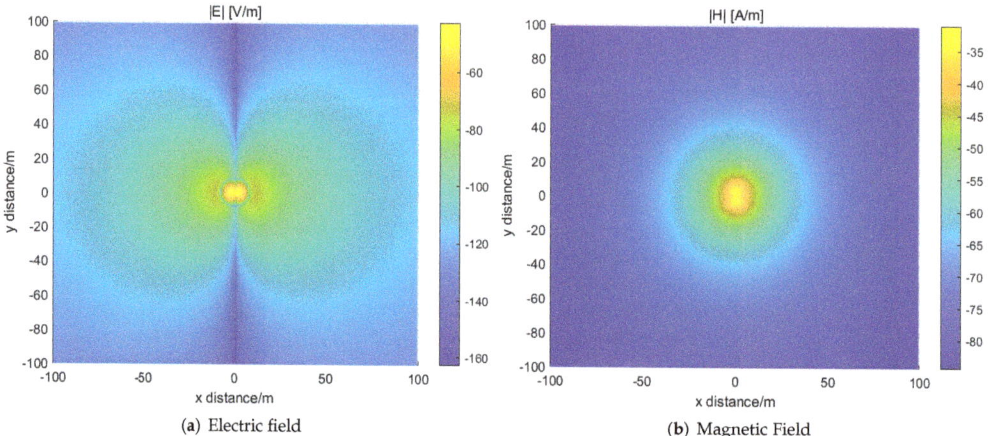

(**a**) Electric field　　　　(**b**) Magnetic Field

Figure 7. The electric field and magnetic field distributions at $xoy - plane$ ($z = 1$ m) of the ELF wave at 5 Hz (HED source located at $z = -1$ m; values are in logarithm form).

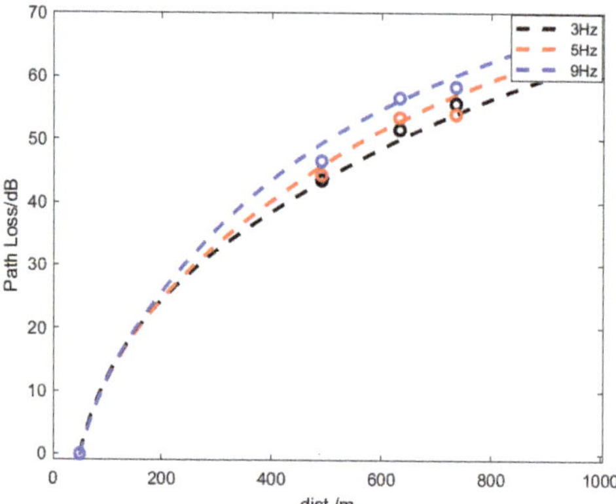

Figure 8. Power loss of the ELF wave at various frequencies at the mountain.

Additionally, a modified Friis Model [25] is applied to model path loss:

$$PL = alg\frac{d}{d_0} + b\frac{d}{d_0} + c \tag{9}$$

where d_0 is 50 m, a is the spreading path loss index and b refers to the absorption index.

The corresponding parameters are shown in Table 2. From the table, it can be easily observed that the spreading path loss index a and absorption index b would be at the same order for different frequencies, as indicated by the figure. The spreading path loss index is around 40, which indicates the appearance of other types of wave transmission in addition to traditional direct propagation, while the absorption index is close to zero because of the non-lossy properties of mountain rocks, which would render absorption loss negligible.

Table 2. Parameters of modified path-loss models.

Freq.	a	b	c	SSE	RMSE
3 Hz	30.86	−0.87	0.46	4.43	0.22
5 Hz	39.56	−0.85	2.11	25.99	0.53
9 Hz	46.17	−0.63	3.43	4.83	0.23

4. Conclusions

In the paper, the emission fields of a submerged line current source in the ground are theoretically and experimentally analyzed. The results showed that the submerged line current source can emit ELF signals and can be modeled as a submerged HED, which would simplify the process for the communication system's design. In addition, the initial studies of the path loss models indicate the potential of the ELF techniques for super long-distance communication, which would pave the way for future studies. However, to render wide ELF bands available and usable, the antenna and transceiver techniques should be fully developed, and they are still in their early phases; thus, more research work should be conducted. As an initial study on such techniques, much work still needs to be performed, such as more experiments and complete theoretical analyses to enrich the database and to explain the peculiarities in detail.

Author Contributions: Conceptualization, K.Y. and S.L.; methodology, K.Y. and J.W.; software, J.W.; validation, J.W. and K.Y.; formal analysis, K.Y. and S.L.; investigation, K.Y. and B.L.; writing—original draft preparation, S.L. and K.Y.; writing—review and editing, K.Y.; visualization, B.L. and K.Y.; supervision, K.Y. and B.L.; project administration, K.Y.; funding acquisition, K.Y., H.L. and K.D. All authors have read and agreed to the published version of the manuscript.

Funding: This research was partly funded by NFSC, grant number 61901386, and partly by Long-Term Funds of Science and Technology on Near-Surface Detection Laboratory, grant number TCGZ2020C002.

Conflicts of Interest: The authors declare no conflict of interest.

References

1. Zenneck, J. Propagation of plane electromagnetic waves along a plane conducting surface and its bearing on the theory of transmission in wireless telegraphy. *Ann. Phys.* **1907**, *23*, 846–866. [CrossRef]
2. Sommerfeld, A. Propagation of waves in wireless telegraphy. *Ann. Phys.* **1909**, *333*, 665–736. [CrossRef]
3. Baños, A.J. *Dipole Radiation in the Presence of a Conducting Half-Space*; Pergamon Press: Oxford, UK, 1966.
4. Pappert, R.A.; Moler, W.F. Propagation theory and Calculations at Lower Extremely Low Frequencis (ELF). *Commun. IEEE Trans.* **1974**, *22*, 438–451. [CrossRef]
5. Chew, W.C. *Waves and Fields in Inhomogeneous Media*; IEEE Press: Piscataway, NJ, USA, 1995.
6. Al-Shamma, A.I.; Shaw, A.; Saman, S. Propagation of electromagnetic waves at MHz frequencies through seawater. *IEEE Trans. Ant. Prop.* **2004**, *52*, 2843–2849. [CrossRef]
7. Wang, H.; Zheng, K.; Yang, K.; Ma, Y. Electromagnetic field in air produced by a horizontal magnetic dipole immersed in sea: Theoretical analysis and experimental results. *IEEE Trans. Ant. Prop.* **2014**, *62*, 4647–4655. [CrossRef]
8. Pan, W.Y.; Li, K. *Propagation of SLF/ELF Electromagnetic Waves*; Springer: Berlin, Germany, 2014.
9. Zhima, Z.; Hu, Y.; Shen, X.; Chu, W.; Piersanti, M.; Parmentier, A.; Zhang, Z.; Wang, Q.; Huang, J.; Zhao, S.; et al. Storm-Time Features of the Ionospheric ELF/VLF Waves and Energetic Electron Fluxes Revealed by the China Seismo-Electromagnetic Satellite. *Appl. Sci.* **2021**, *11*, 2617. [CrossRef]
10. Chave, A.D.; Flosadottir, A.; Cox, C.S. Some Comments on Seabed Propagation of ULF/ELF Electromagnetic fields. *Radio Sci.* **2016**, *25*, 825–836. [CrossRef]
11. Wang, J.; Li, B. Electromagnetic Fields Generated above a Shallow Sea by a Submerged Horizontal Electric Dipole. *IEEE Trans. Antenna Propag.* **2017**, *65*, 2707–2712. [CrossRef]
12. Xiang, G.; Yan, S.; Li, B. Localisation of Ferromagnetic Target with Three Magnetic Sensors in the Movement Considering Angular Rotation. *Sensors* **2017**, *17*, 2079.
13. Wang, J.; Li, B.; Chen, L.; Li, L. A Novel Detection Method for Underwater Moving Targets by Measuring Their ELF Emissions with Inductive Sensors. *Sensors* **2017**, *17*, 1734. [CrossRef] [PubMed]
14. He, T.; He, L.N.; Li, K. Poynting vector of an ELF electromagnetic wave in three-layered ocean floor. *J. Electromagn. Waves Appl.* **2018**, *32*, 2339–2349. [CrossRef]
15. Li, F.; Yang, Z.; Fan, Y.; Li, Y.; Li, G. Application of a New Combination Algorithm in ELF-EM Processing. *Symmetry* **2020**, *12*, 337. [CrossRef]
16. King, R.W.P.; Wu, T.T. Lateral wave: Formulas for the magnetic field. *J. Appl. Phys.* **1983**, *54*, 507–514. [CrossRef]
17. King, R.W.P. Electromagnetic surface waves: New formulas and their application to determine the electrical properties of the sea bottom. *J. Appl. Phys.* **1985**, *58*, 3612–3624. [CrossRef]
18. King, R.W.P.; Wu, T.T. Properties of lateral electromagnetic fields and their application. *Radio Sci.* **1986**, *21*, 13–23. [CrossRef]
19. Dunn, J.M. Lateral wave propagation in a three-layered medium. *Radio Sci.* **1986**, *21*, 787–796. [CrossRef]
20. Bogie, S. Conduction and Magnetic Signaling in the Sea. *Radio Electron. Eng.* **1972**, *42*, 447–452. [CrossRef]

21. Diegel, M.; King, R.W.P. 1973: Electromagnetic propagation between antennas submerged in the ocean. *IEEE Trans. Ant. Prop.* **1973**, *4*, 507–513 [CrossRef]
22. Hoh, Y.S. Propagation Prediction for the Navy Extremely Low Frequency Program. *IEEE J. Ocean. Eng.* **1984**, *9*, 164–178.
23. King, R.W.P.; Owen, M.; Wu, T.T. *Lateral Electromagnteic Waves: Theory and Applications to Communications, Geophysical Exporation, and Remote Sensing*; Springer: New York, NY, USA, 1992.
24. Li, K. *Electromagnetic Fields in Stratified Media*; Springer: Berlin, Germany, 2009.
25. Yang, K.; Pellegrini, A.; Munoz, M.; Brizzi, A.; Alomainy, A.; Hao, Y. Numerical Analysis and Characterisation of THz Propagation Channel for Body-Centric Nano-Communications. *IEEE Trans. Thz Sci. Technol.* **2015**, *5*, 419–426.

MDPI

St. Alban-Anlage 66

4052 Basel

Switzerland

www.mdpi.com

Electronics Editorial Office

E-mail: electronics@mdpi.com

www.mdpi.com/journal/electronics